D0820973

STUDENT MATHEMATICAL LIBRARY
Ω IAS/PARK CITY MATHEMATICAL SUBSERIES
Volume 33

Lectures in Geometric Combinatorics

Rekha R. Thomas

American Mathematical Society
Institute for Advanced Study

2000 *Mathematics Subject Classification*. Primary 52–01;
Secondary 13–01.

For additional information and updates on this book, visit
www.ams.org/bookpages/stml-33

Library of Congress Cataloging-in-Publication Data

Thomas, Rekha R., 1967–
    Lectures in geometric combinatorics / Rekha R. Thomas.
        p. cm. — (Student mathematical library, ISSN 1520-9121 ; v. 33. IAS/Park City mathematical subseries)
    Includes bibliographical references and index.
    ISBN 0-8218-4140-8 (alk. paper)
    1. Combinatorial geometry.   2. Combinatorial analysis.   I. Title.   II. Series: Student mathematical library ; v. 33.   III. Series: Student mathematical library. IAS/Park City mathematical subseries.

QA167.T45   2006
516′.13—dc22   2006042841

Contents

IAS/Park City
Mathematics Institute

The IAS/Park City Mathematics Institute (PCMI) was founded in 1991 as part of the "Regional Geometry Institute" initiative of the National Science Foundation. In mid-1993 the program found an institutional home at the Institute for Advanced Study (IAS) in Princeton, New Jersey. The PCMI continues to hold summer programs in Park City, Utah.

The IAS/Park City Mathematics Institute encourages both research and education in mathematics and fosters interaction between the two. The three-week summer institute offers programs for researchers and postdoctoral scholars, graduate students, undergraduate students, high school teachers, mathematics education researchers, and undergraduate faculty. One of PCMI's main goals is to make all of the participants aware of the total spectrum of activities that occur in mathematics education and research: we wish to involve professional mathematicians in education and to bring modern concepts in mathematics to the attention of educators. To that end the summer institute features general sessions designed to encourage interaction among the various groups. In-year activities at sites around the country form an integral part of the High School Teacher Program.

Each summer a different topic is chosen as the focus of the Research Program and Graduate Summer School. Activities in the Undergraduate Program deal with this topic as well. Lecture notes from the Graduate Summer School are published each year in the IAS/Park City Mathematics Series. Course materials from the Undergraduate Program, such as the current volume, are now being published as part of the IAS/Park City Mathematical Subseries in the Student Mathematical Library. We are happy to make available more of the excellent resources which have been developed as part of the PCMI.

John Polking, Series Editor
February 20, 2006

Preface

These lectures were prepared for the advanced undergraduate course in *Geometric Combinatorics* at the Park City Mathematics Institute in July 2004. Many thanks to the organizers of the undergraduate program, Bill Barker and Roger Howe, for inviting me to teach this course. I also wish to thank Ezra Miller, Vic Reiner and Bernd Sturmfels, who coordinated the graduate research program at PCMI, for their support. Edwin O'Shea conducted all the tutorials at the course and wrote several of the exercises seen in these lectures. Edwin was a huge help in the preparation of these lectures from beginning to end.

The main goal of these lectures was to develop the theory of convex polytopes from a geometric viewpoint to lead up to recent developments centered around secondary and state polytopes arising from point configurations. The geometric viewpoint naturally relies on linear optimization over polytopes. Chapters 2 and 3 develop the basics of polytope theory. In Chapters 4 and 5 we see the tools of Schlegel and Gale diagrams for visualizing polytopes and understanding their facial structure. Gale diagrams have been used to unearth several bizarre phenomena in polytopes, such as the existence of polytopes whose vertices cannot have rational coordinates and others whose facets cannot be prescribed. These examples are described in Chapter 6. In Chapters 7–9 we construct the secondary polytope of a graded point configuration. The faces of this polytope index the regular subdivisions of the configuration. Secondary polytopes appeared in the literature in the early 1990's and play a crucial role in combinatorics, discrete optimization and algebraic geometry. The secondary polytope of a point configuration is naturally refined by the state polytope of the toric ideal of the configuration. In Chapters 10–14 we establish this relationship. The state polytope of a toric ideal arises from the theory of Gröbner bases, which is developed in Chapters 10–12. Chapter 13 establishes the connection between the Gröbner bases of a toric ideal and the regular triangulations of the point configuration defining the ideal. Finally, in Chapter 14 we construct the state polytope of a toric ideal and relate it to the corresponding secondary polytope.

These lectures are meant to be self-contained and do not require any background beyond basic linear algebra. The concepts needed from abstract algebra are developed in Chapters 1, 10, 11 and 12.

I wish to thank Tristram Bogart, Ezra Miller, Edwin O'Shea and Alex Papazoglu for carefully proofreading many parts of the original manuscript. Ezra made several important remarks and corrections that have greatly benefited this final version. Many thanks also to Sergei Gelfand and Ed Dunne at the AMS office for their patience and help in publishing this book. Lastly, I wish to thank Peter Blossey for twenty-four hour technical assistance in preparing this book.

The author was supported in part by grants DMS-0100141 and DMS-0401047 from the National Science Foundation.

Rekha R. Thomas
Seattle, January 2006

Chapter 1

Abstract Algebra: Groups, Rings and Fields

This course will aim at understanding *convex polytopes*, which are fundamental geometric objects in combinatorics, using techniques from algebra and discrete geometry. Polytopes arise everywhere in the real world and in mathematics. The most famous examples are the Platonic solids in three-dimensional space: *cube, tetrahedron, octahedron, icosahedron* and *dodecahedron*, which were known to the ancient Greeks. The natural first approach to understanding polytopes should be through geometry as they are first and foremost geometric objects. However, any experience with visualizing geometric objects will tell you soon that geometry is already quite hard in three-dimensional space, and if one has to study objects in four- or higher-dimensional space, then it is essentially hopeless to rely only on our geometric and drawing skills. This frustration led mathematicians to the discovery that algebra can be used to encode geometry and, since algebra does not suffer from the same limitations as geometry in dealing with higher dimensions, it can serve very well as the language of geometry. A simple example of this translation can be seen by noting that, while it is hard to visualize vectors in four-dimensional space, linear algebra allows us to work with their algebraic incarnations $\mathbf{v} = (v_1, v_2, v_3, v_4) \in \mathbb{R}^4$ and $\mathbf{w} = (w_1, w_2, w_3, w_4) \in \mathbb{R}^4$ and to manipulate them to find new quantities, such as the sum vector $\mathbf{v} + \mathbf{w} = (v_1 + w_1, v_2 + w_2, v_3 + w_3, v_4 + w_4) \in \mathbb{R}^4$ or the length of their difference vector $\sqrt{(v_1 - w_1)^2 + \cdots + (v_4 - w_4)^2}$. We use $\mathbb{R}$ for the set of real numbers.

These lectures will focus on techniques from linear and abstract algebra to understand the geometry and combinatorics of polytopes. We begin with some basic abstract algebra. The algebraically sophisticated reader should skip ahead to the next chapter and refer back to this chapter only as needed. The material in this lecture is taken largely from the book [**DF91**].

In linear algebra one learns about *vector spaces* over *fields*. Both of these objects are examples of a more basic object known as a *group*.

Definition 1.1. A set G along with an operation $*$ on pairs of elements of G is called a **group** if the pair $(G, *)$ satisfies the following properties:

(1) $*$ *is a binary operation on G*: This means that for any two elements $g_1, g_2 \in G$, $g_1 * g_2 \in G$. In other words, G is *closed* under the operation $*$ on its elements.

(2) $*$ *is associative*: For any three elements $g_1, g_2, g_3 \in G$, $(g_1 * g_2) * g_3 = g_1 * (g_2 * g_3)$.

(3) *G has an identity element with respect to $*$*: This means that there is an element $e \in G$ such that for all $g \in G$, $e * g = g * e = g$. If $*$ is addition, then e is usually written as 0. If $*$ is multiplication, then e is usually written as 1.

(4) *Every $g \in G$ has an inverse*: For each $g \in G$ there is an element $g^{-1} \in G$ such that $g * g^{-1} = g^{-1} * g = e$. If $*$ is addition, then it is usual to write g^{-1} as $-g$.

It can be proved that the identity element in G is unique and that every element in G has a unique inverse. Let $\mathbb{Z}$ be the set of integers and let $\mathbb{R}^* := \mathbb{R} \backslash \{0\}$. The **multiplication table** of a finite group is a $|G| \times |G|$ array whose rows and columns are indexed by the elements of G and the entry in the box with row index g and column index g' is the product $g * g'$.

Exercise 1.2. Check that the following are groups. In each case, write down how the binary operation works, the identity element of the group, and the inverse of an arbitrary element in the group.

(1) $(\mathbb{Z}^n, +)$

(2) $(\mathbb{R}^*, \times)$

(3) $((\mathbb{R}^*)^n, \times)$

The above groups are all infinite. We now study two important families of finite groups that are useful in the study of polytopes.

The symmetric group S_n:
Recall that a **permutation** of n letters $1, 2, \ldots, n$ is any arrangement of the n letters or, more formally, a one-to-one onto function from the set $[n] := \{1, 2, \ldots, n\}$ to itself. Permutations are denoted by the small Greek letters σ, τ, etc. and they can be written in many ways. For instance, the permutation

$$\sigma : \{1, 2, 3\} \rightarrow \{1, 2, 3\} : 1 \mapsto 2,\ 2 \mapsto 1,\ 3 \mapsto 3$$

is denoted as either $\begin{pmatrix} 1 & 2 & 3 \\ 2 & 1 & 3 \end{pmatrix}$ or, more compactly, by recording just the last row as 213. Since permutations are functions, two permutations can be composed in the usual way that functions are composed: $f \circ g$ is the function obtained by first applying g and then applying f. The symbol $\circ$ denotes composition. Check that $213 \circ 321 = 312$, which is again a permutation. Let S_n denote the set of all permutations on n letters. Then $(S_n, \circ)$ is a group with $n!$ elements. We sometimes say that 312 is the *product* $213 \circ 321$.

Exercise 1.3. (1) Check that $(S_n, \circ)$ is a group for any positive integer n. What is the identity element of this group, and what is the inverse of a permutation $\sigma \in S_n$?

(2) List the elements of S_2 and S_3, and compute their multiplication tables.

Definition 1.4. The group $(G, *)$ is **abelian** if for all $g, g' \in G$, $g * g' = g' * g$.

Check that S_3 is not an abelian group. Do you see how to use this to prove that $(S_n, \circ)$ is not abelian for all $n \geq 3$?

We now study a second family of non-abelian groups. The **regular n-gon**, which is a polygon with n sides of equal length, is an example of a *polytope* in $\mathbb{R}^2$. *Regular* polygons have all sides of equal length and the same angle between any two adjacent sides. For instance, an equilateral triangle is a regular 3-gon, a square is a regular 4-gon, a pentagon with equal sides and angles is a regular 5-gon, etc.

The dihedral group D_{2n}:

The group D_{2n} is the *group of symmetries* of a regular n-gon. A **symmetry** of a regular n-gon is any rigid motion obtained by taking a copy of the n-gon, moving this copy in any fashion in three-dimensional space and placing it back down so that the copy exactly covers the original n-gon. Mathematically, we can describe a symmetry s by a permutation in S_n. Fix a cyclic labeling of the corners (*vertices*) of the n-gon by the letters $1, 2, \ldots, n$. If s puts vertex i in the place where vertex j was originally, then the permutation s sends i to j. Note that since our labeling was cyclic, s is completely specified by noting where the vertices 1 and 2 are sent. In particular, this implies that s cannot be any permutation in S_n.

How many symmetries are there for a regular n-gon? Given a vertex i, there is a symmetry that sends vertex 1 to i. Then vertex 2 has to go to either vertex $i - 1$ or vertex $i + 1$. Note that we have to add modulo n and hence $n + 1$ is 1 and $1 - 1$ is n. By following the first symmetry by a reflection of the n-gon about the line joining the center of the n-gon to vertex i, we see that there are symmetries that send 2 to either $i - 1$ or $i + 1$. Thus there are $2n$ positions that the ordered pair of vertices 1 and 2 may be sent to by symmetries. However, since every symmetry is completely determined by what happens to 1 and 2, we conclude that there are exactly $2n$ symmetries of the regular n-gon. These $2n$ symmetries are the n *rotations* about the center through $\frac{2\pi i}{n}$ radians for $1 \leq i \leq n$ and the n *reflections* through the n lines of symmetry. If n is odd, each symmetry line passes through a vertex and the midpoint of the opposite side. If n is even, there are $n/2$ lines of symmetry which pass through two opposite vertices and $n/2$ which perpendicularly bisect two opposite sides. The dihedral group D_{2n} is the set of all symmetries of a regular n-gon with the binary operation of composition of symmetries (which are permutations).

Example 1.5. Let $\square$ be a square with vertices $1, 2, 3, 4$ that are labeled counterclockwise from the bottom left vertex and centered about the origin in $\mathbb{R}^2$. Then its group of symmetries is the dihedral group

$$D_8 = \left\{ \begin{array}{ll} e = \begin{pmatrix} 1 & 2 & 3 & 4 \\ 1 & 2 & 3 & 4 \end{pmatrix}, & s = \begin{pmatrix} 1 & 2 & 3 & 4 \\ 4 & 3 & 2 & 1 \end{pmatrix}, \\ r = \begin{pmatrix} 1 & 2 & 3 & 4 \\ 2 & 3 & 4 & 1 \end{pmatrix}, & r^2 = \begin{pmatrix} 1 & 2 & 3 & 4 \\ 3 & 4 & 1 & 2 \end{pmatrix}, \\ r^3 = \begin{pmatrix} 1 & 2 & 3 & 4 \\ 4 & 1 & 2 & 3 \end{pmatrix}, & r \circ s = \begin{pmatrix} 1 & 2 & 3 & 4 \\ 1 & 4 & 3 & 2 \end{pmatrix}, \\ r^2 \circ s = \begin{pmatrix} 1 & 2 & 3 & 4 \\ 2 & 1 & 4 & 3 \end{pmatrix}, & r^3 \circ s = \begin{pmatrix} 1 & 2 & 3 & 4 \\ 3 & 2 & 1 & 4 \end{pmatrix} \end{array} \right\}$$

where r denotes counterclockwise rotation by 90 degrees about the origin and s denotes reflection about the horizontal axis.

Exercise 1.6. Fix a labeling of a regular n-gon (say counterclockwise, starting at some vertex). Let r denote counterclockwise rotation through $\frac{2\pi}{n}$ radians and let s denote reflection about the line of symmetry through the center of the n-gon and vertex 1. Then show the following.

(1) $D_{2n} = \{e, r, r^2, \ldots, r^{n-1}, s, sr, sr^2, \ldots, sr^{n-1}\}$.

(2) What are the inverses in the above group?

(**Hint:** (i) $1, r, r^2, \ldots, r^{n-1}$ are all distinct, (ii) $r^n = e$, (iii) $s^2 = e$, (iv) $s \neq r^i$ for any i, (v) $sr^i \neq sr^j$ for all $0 \leq i, j \leq n - 1$, $i \neq j$, (vi) $sr = r^{-1}s$, (vii) $sr^i = r^{-i}s$, for $0 \leq i \leq n$.)

Exercise 1.7. Let G be the symmetries of a regular cube in $\mathbb{R}^3$. Show that $|G| = 24$.

Definition 1.8. A set R with two binary operations $+$ and $\times$ is called a **ring** if the following conditions are satisfied:

(1) $(R, +)$ is an abelian group,

(2) $\times$ is associative : $(a \times b) \times c = a \times (b \times c)$ for all $a, b, c \in R$,

(3) $\times$ distributes over $+$: for all $a, b, c \in R$,

$$(a + b) \times c = (a \times c) + (b \times c), \text{ and } a \times (b + c) = (a \times b) + (a \times c).$$

If, in addition, R has an identity element with respect to $\times$, we say that R is a ring with identity. If $\times$ is commutative in R, then we say that R is a commutative ring. The identity of $(R, +)$ is the additive identity in R denoted as 0 while the multiplicative identity, if it exists, is denoted as 1. We will only consider commutative rings with identity.

Exercise 1.9.   (1) Show that $(\mathbb{Z}, +, \times)$ is a commutative ring with identity.

(2) Let M_n denote the set of $n \times n$ matrices with entries in $\mathbb{R}$. Then show that under the usual operations of matrix addition and multiplication, M_n is a non-commutative ring with identity. Is $(M_n, \times)$ a group?

Definition 1.10. A **field** is a set F with two binary operations $+$ and $\times$ such that both $(F, +)$ and $(F^* := F \backslash \{0\}, \times)$ are abelian groups and the following distributive law holds:

$$a \times (b + c) = (a \times b) + (a \times c), \text{ for all } a, b, c \in F.$$

Let $\mathbb{C}$ denote the set of complex numbers and let $\mathbb{Q}$ denote the set of rational numbers.

Exercise 1.11. Check that $(\mathbb{C}, +, \times)$, $(\mathbb{R}, +, \times)$, $(\mathbb{Q}, +, \times)$ are fields while $(\mathbb{Z}, +, \times)$ and $(M_n, +, \times)$ are not fields.

Where does a vector space fit in the above hierarchy?

Chapter 2

Convex Polytopes: Definitions and Examples

In this chapter we define the notion of a convex polytope. There are several excellent books on polytopes. Much of the material on polytopes in this book is taken from [**Grü03**] and [**Zie95**]. We start with an example of a family of convex polytopes.

Example 2.1. Cubes: The following is an example of the familiar three-dimensional cube:

$$C_3 := \left\{ (x_1, x_2, x_3) \in \mathbb{R}^3 : \begin{array}{l} 0 \leq x_1 \leq 1 \\ 0 \leq x_2 \leq 1 \\ 0 \leq x_3 \leq 1 \end{array} \right\}.$$

The cube C_3 has volume one and edges of length one. By translating this cube around in $\mathbb{R}^3$, we see that there are infinitely many three-dimensional cubes (3-cubes) of volume one and edges of length one in $\mathbb{R}^3$. If you are interested in studying the properties of these cubes, you might be willing to believe that it suffices to examine one member in this infinite family. Thus we pick the above member of the family and call it *the* three-dimensional **unit cube**.

The unit 3-cube is of course the older sibling of a square in $\mathbb{R}^2$. Again, picking a representative, we have *the* unit square (or the unit

2-cube):

$$C_2 := \left\{ (x_1, x_2) \in \mathbb{R}^2 : \begin{array}{c} 0 \le x_1 \le 1 \\ 0 \le x_2 \le 1 \end{array} \right\}.$$

Going down in the family, we could ask who the 1-cube is. If we simply mimic the pattern, we might conclude that the unit 1-cube is the line segment:

$$C_1 := \left\{ (x_1) \in \mathbb{R} : \ 0 \le x_1 \le 1 \ \right\}.$$

The baby of the family is the 0-cube $C_0 = \{0\} = \mathbb{R}^0$.

How about going up in the family? What might be the unit 4-cube? Again, simply mimicking the pattern, we might define it to be

$$C_4 := \left\{ (x_1, x_2, x_3, x_4) \in \mathbb{R}^4 : \begin{array}{c} 0 \le x_1 \le 1 \\ 0 \le x_2 \le 1 \\ 0 \le x_3 \le 1 \\ 0 \le x_4 \le 1 \end{array} \right\}.$$

Of course this is hard to visualize. In Chapter 4 we will learn about Schlegel diagrams that can be used to see C_4. Making life even harder, we could define the unit d-cube (the unit cube of dimension d) to be

$$C_d := \left\{ (x_1, \ldots, x_d) \in \mathbb{R}^d : \begin{array}{c} 0 \le x_1 \le 1 \\ 0 \le x_2 \le 1 \\ \vdots \\ 0 \le x_d \le 1 \end{array} \right\},$$

thus creating an infinite family of unit cubes $\{C_d : d \in \mathbb{N}\}$. The symbol $\mathbb{N}$ denotes the set of non-negative integers $\{0, 1, 2, 3, \ldots\}$. Every member of this family is a convex polytope.

Definition 2.2. A set $C \subseteq \mathbb{R}^d$ is **convex** if for any two points $\mathbf{p}$ and $\mathbf{q}$ in C, the entire line segment joining them, $\{\lambda\mathbf{p} + (1 - \lambda)\mathbf{q} : 0 \le \lambda \le 1\}$, is contained in C.

Exercise 2.3. (1) Check that each C_d, $d \in \mathbb{N}$, is convex.
(2) Draw an example of a non-convex set.

Recall from linear algebra that a **hyperplane** in $\mathbb{R}^d$ is a set

$$H := \{\mathbf{x} \in \mathbb{R}^d : a_1x_1 + a_2x_2 + \cdots + a_dx_d = b\}$$

where $a_1, \ldots, a_d, b \in \mathbb{R}$. The hyperplane H defines two **halfspaces** in $\mathbb{R}^d$:

$$H^+ := \{\mathbf{x} \in \mathbb{R}^d : a_1x_1 + a_2x_2 + \cdots + a_dx_d \geq b\} \text{ and}$$
$$H^- := \{\mathbf{x} \in \mathbb{R}^d : a_1x_1 + a_2x_2 + \cdots + a_dx_d \leq b\}.$$

We can always write H^+ with a $\leq$ sign by simply multiplying both sides of the inequality by -1 to get

$$H^+ = \{\mathbf{x} \in \mathbb{R}^d : -a_1x_1 - a_2x_2 - \cdots - a_dx_d \leq -b\}.$$

Thus all halfspaces can be considered to be of the form H^-. The common intersection of several halfspaces then looks like

$$\left\{ \mathbf{x} \in \mathbb{R}^d : \begin{array}{c} a_{11}x_1 + a_{12}x_2 + \cdots + a_{1d}x_d \leq b_1 \\ a_{21}x_1 + a_{22}x_2 + \cdots + a_{2d}x_d \leq b_2 \\ \vdots \\ a_{m1}x_1 + a_{m2}x_2 + \cdots + a_{md}x_d \leq b_m \end{array} \right\}$$

which can be written more compactly as $\{\mathbf{x} \in \mathbb{R}^d : A\mathbf{x} \leq \mathbf{b}\}$ where $\mathbf{b} = (b_1, \ldots, b_m) \in \mathbb{R}^m$ and A is the $m \times d$ matrix

$$A = \begin{pmatrix} a_{11} & a_{12} & \cdots & a_{1d} \\ \vdots & \vdots & \vdots & \vdots \\ a_{m1} & a_{m2} & \cdots & a_{md} \end{pmatrix} =: (a_{ij})_{m \times d}.$$

Exercise 2.4. Check that the d-cube C_d is the common intersection of $2d$ halfspaces in $\mathbb{R}^d$.

Definition 2.5. A **polyhedron** (or $\mathcal{H}$-**polyhedron**) in $\mathbb{R}^d$ is any set obtained as the intersection of finitely many halfspaces in $\mathbb{R}^d$. Mathematically, it has the form $P = \{\mathbf{x} \in \mathbb{R}^d : A\mathbf{x} \leq \mathbf{b}\}$ where $A \in \mathbb{R}^{m \times d}$ and $\mathbf{b} \in \mathbb{R}^m$.

The prefix $\mathcal{H}$ stands for the fact that the above definition of a polyhedron involves intersecting halfspaces. We say that a set is bounded if we can enclose it entirely in a ball of some finite radius. Note that all our unit cubes are bounded.

Definition 2.6. A bounded $\mathcal{H}$-polyhedron is called an $\mathcal{H}$-**polytope**.

Example 2.7. We can create unbounded polyhedra by throwing away some of the inequalities from our d-cubes. For instance if we take no inequalities, then we get $P = \mathbb{R}^d$ which is an unbounded polyhedron. Any single halfspace in $\mathbb{R}^d$, $d \geq 1$, is also an unbounded polyhedron. For example, $\{x \in \mathbb{R} : x \geq 0\}$ and $\{x \in \mathbb{R} : x \leq 1\}$ are examples of halfspaces in $\mathbb{R}$. Both are unbounded polyhedra obtained by *relaxing* inequalities of C_1.

Lemma 2.8. *An $\mathcal{H}$-polyhedron P is a convex set.*

Proof. Consider two points $\mathbf{p}, \mathbf{q} \in P$ and any point $\lambda \mathbf{p} + (1 - \lambda)\mathbf{q}$ on the line segment joining them. Then $A(\lambda \mathbf{p} + (1 - \lambda)\mathbf{q}) = \lambda A\mathbf{p} + (1 - \lambda)A\mathbf{q} \leq \lambda \mathbf{b} + (1 - \lambda)\mathbf{b} = \mathbf{b}$ where the inequality follows from the fact that $0 \leq \lambda \leq 1$, which makes $0 \leq 1 - \lambda \leq 1$. $\square$

There are objects called *non-convex polyhedra* which we will not touch in these chapters. We will only consider convex polyhedra and, in fact, mostly only convex polytopes, and we will drop the adjective *convex* from now on. There is a second equivalent way to define polytopes which makes their convexity more explicit.

Definition 2.9. A **convex combination** of any two points $\mathbf{p}, \mathbf{q} \in \mathbb{R}^d$ is any point of the form $\lambda \mathbf{p} + (1 - \lambda)\mathbf{q}$ where $0 \leq \lambda \leq 1$. The set of all convex combinations of $\mathbf{p}$ and $\mathbf{q}$ is called the **convex hull** of $\mathbf{p}$ and $\mathbf{q}$.

Remark 2.10. The convex hull of $\mathbf{p}$ and $\mathbf{q}$ is the line segment joining them.

Definition 2.11. A **convex combination** of $\mathbf{p}_1, \ldots, \mathbf{p}_t \in \mathbb{R}^d$ is any point of the form $\sum_{i=1}^{t} \lambda_i \mathbf{p}_i$ where $\lambda_i \geq 0$ for all $i = 1, \ldots, t$ and $\sum_{i=1}^{t} \lambda_i = 1$. The set of all convex combinations of $\mathbf{p}_1, \ldots, \mathbf{p}_t$ is called their **convex hull**. We denote it as $\mathrm{conv}(\{\mathbf{p}_1, \ldots, \mathbf{p}_t\})$.

Exercise 2.12. (1) Draw the convex hull of the points 0 and 1 in $\mathbb{R}$.
(2) Draw the convex hull of $(0, 0), (1, 0), (0, 1), (1, 1)$ in $\mathbb{R}^2$.
(3) What is C_3 the convex hull of? How about C_d?

Taking the convex hull of a finite set of points is like "shrink wrapping" the points. Here are some basic facts about convex hulls:

- $\mathrm{conv}(\{\mathbf{p}_1, \ldots, \mathbf{p}_t\})$ is convex,

- $\text{conv}(\{\mathbf{p}_1, \ldots, \mathbf{p}_t\})$ is the smallest convex set containing $\mathbf{p}_1, \ldots, \mathbf{p}_t$,

- $\text{conv}(\{\mathbf{p}_1, \ldots, \mathbf{p}_t\})$ is the intersection of all convex sets containing $\mathbf{p}_1, \ldots, \mathbf{p}_t$. This follows from the fact that the intersection of two convex sets is convex.

Definition 2.13. A $\mathcal{V}$-**polytope** in $\mathbb{R}^d$ is the convex hull of a finite number of points in $\mathbb{R}^d$. Mathematically, it is a set of the form $P = \text{conv}(\{\mathbf{p}_1, \ldots, \mathbf{p}_t\})$ where $\mathbf{p}_1, \ldots, \mathbf{p}_t \in \mathbb{R}^d$.

The prefix $\mathcal{V}$ stands for the polytope being expressed as the convex hull of a set of points $\mathcal{V}$. It takes some serious work to prove that the two notions of a polytope in Definitions 2.5 and 2.13 are the same.

Theorem 2.14. Main Theorem of Polytope Theory [Zie95, Theorem 1.1]. *Every $\mathcal{V}$-polytope has a description by inequalities as an $\mathcal{H}$-polytope and every $\mathcal{H}$-polytope is the convex hull of a minimal number of finitely many points called its* **vertices**.

The process of converting one description to another is called *Fourier-Motzkin elimination.* See [**Zie95**, §1.2] for instance.

Example 2.15. The **cross-polytope** in $\mathbb{R}^d$ is the $\mathcal{V}$-polytope

$$C_d^{\Delta} := \text{conv}(\{\pm\mathbf{e}_1, \pm\mathbf{e}_2, \ldots, \pm\mathbf{e}_d\})$$

where $\mathbf{e}_1, \ldots, \mathbf{e}_d$ are the standard unit vectors in $\mathbb{R}^d$. Note that C_1^{Δ} is a line segment in $\mathbb{R}$, C_2^{Δ} is a "diamond" in $\mathbb{R}^2$ and C_3^{Δ} is an *octahedron.* Expressed as an $\mathcal{H}$-polytope,

$$C_3^{\Delta} = \left\{ (x_1, x_2, x_3) \in \mathbb{R}^3 : \begin{array}{rcl} x_1 + x_2 + x_3 & \leq & 1 \\ -x_1 + x_2 + x_3 & \leq & 1 \\ x_1 - x_2 + x_3 & \leq & 1 \\ x_1 + x_2 - x_3 & \leq & 1 \\ -x_1 - x_2 + x_3 & \leq & 1 \\ -x_1 + x_2 - x_3 & \leq & 1 \\ x_1 - x_2 - x_3 & \leq & 1 \\ -x_1 - x_2 - x_3 & \leq & 1 \end{array} \right\}.$$

Exercise 2.16. What is the $\mathcal{H}$-representation of C_d^{Δ}?

Example 2.17. Can you devise an algorithm for converting the $\mathcal{H}$-representation of a polytope to its $\mathcal{V}$-representation and vice versa?

Example 2.18. The simplest polytopes are the **simplices**. The unit $(d-1)$-simplex is defined as

$$\begin{aligned}
\Delta_{d-1} \; &:= \; \operatorname{conv}(\{\mathbf{e}_1, \mathbf{e}_2, \dots, \mathbf{e}_d\}) \\
&= \; \{\mathbf{x} \in \mathbb{R}^d \, : \, x_1 + \cdots + x_d = 1, x_i \geq 0, \; i = 1, \dots, d\}.
\end{aligned}$$

This family includes a line segment, triangle and tetrahedron.

Exercise 2.19. Show that if you drop any inequality in the $\mathcal{H}$-description of Δ_{d-1}, you get an unbounded polyhedron.

The $(d-1)$-simplex Δ_{d-1} lives entirely in the hyperplane $\{\mathbf{x} \in \mathbb{R}^d \, : \, x_1 + \cdots + x_d = 1\}$ and hence is a $(d-1)$-dimensional polytope even though it lives in $\mathbb{R}^d$. You can verify this at least for $d = 1, 2, 3$. The *dimension* of a polytope is perhaps its most important invariant. It is what we think it is in low-dimensional spaces, but we need a formal definition so that we can calculate it for polytopes that live in high-dimensional spaces. To do this, we need to understand affine spaces, which are closely related to linear vector spaces.

Definition 2.20.   (1) An **affine combination** of the vectors $\mathbf{p}_1, \dots, \mathbf{p}_t \in \mathbb{R}^d$ is a combination of the form $\sum_{i=1}^{t} \lambda_i \mathbf{p}_i$ such that $\sum_{i=1}^{t} \lambda_i = 1$.

(2) The **affine hull** of $\mathbf{p}_1, \dots, \mathbf{p}_t$ is the set of all affine combinations of $\mathbf{p}_1, \dots, \mathbf{p}_t$. It is denoted as $\operatorname{aff}(\{\mathbf{p}_1, \dots, \mathbf{p}_t\})$.

(3) The affine hull of a set S, denoted as $\operatorname{aff}(S)$, is the set of all affine combinations of finitely many points in S.

Example 2.21. (1) The affine hull of two points $\mathbf{p}$ and $\mathbf{q}$ in $\mathbb{R}^d$ is the line through them. Note the difference between $\operatorname{aff}(\{\mathbf{p}, \mathbf{q}\})$ and $\operatorname{conv}(\{\mathbf{p}, \mathbf{q}\})$.

(2) The affine hull of the three points $(-1, 1), (0, 1), (1, 1) \in \mathbb{R}^2$ is the line $x_2 = 1$ which also equals the affine hull of the polytope $\operatorname{conv}(\{(-1, 1), (0, 1), (1, 1)\})$.

(3) The affine hull of the cube $C_3 \subset \mathbb{R}^3$ is $\mathbb{R}^3$.

Definition 2.22. An **affine space** is any set of the form $\{\mathbf{x} \in \mathbb{R}^d \, : \, A\mathbf{x} = \mathbf{b}\}$.

A non-empty affine space $\{\mathbf{x} \in \mathbb{R}^d : A\mathbf{x} = \mathbf{b}\}$ is a translate of the vector space $\{\mathbf{x} \in \mathbb{R}^d : A\mathbf{x} = \mathbf{0}\}$. Furthermore, every affine hull is an affine space. In (2) above, $\mathrm{aff}(\{(-1,1),(0,1),(1,1)\}) = \{(x_1, x_2) : x_2 = 1\}$ is the translation of the vector space $\{(x_1, x_2) : x_2 = 0\}$. In (3) above, $\mathrm{aff}(C_3)$ is already the linear vector space $\mathbb{R}^3$.

Definition 2.23. (1) The **dimension** of an affine space is the dimension of the linear vector space that it is a translate of.

(2) The **dimension** of a polytope P is the dimension of its affine hull.

Exercise 2.24. Calculate the dimension of the polytope

$$\mathrm{conv}(\{(1,0,0),(1,1,0),(1,2,1),(1,1,2),(1,0,1)\})$$

in $\mathbb{R}^3$. Based on your answer, how many *equations* do you expect to see in the $\mathcal{H}$-representation of this polytope?

Let P be a d-dimensional polytope (d-polytope) in $\mathbb{R}^d$ containing the origin in its interior. Let $\mathbf{e}_i \in \mathbb{R}^d$ be the ith standard unit vector in $\mathbb{R}^d$. The three most basic constructions of polytopes with P as their foundation are the *pyramid*, *prism* and *bipyramid* over P.

Pyramid: This is the convex hull of P and any point outside its affine hull. To construct one example, embed P in $\mathbb{R}^{d+1}$ as $\{(\mathbf{p}, 0) : \mathbf{p} \in P\}$. Then $\mathrm{pyr}(P) := \mathrm{conv}(P, \mathbf{e}_{d+1})$. The much visited convex polytope, the Egyptian pyramid, is an example of $\mathrm{pyr}(C_2)$. The d-simplex Δ_d equals $\mathrm{pyr}(\Delta_{d-1}) = \mathrm{pyr}(\mathrm{pyr}(\Delta_{d-2})) = \cdots = \mathrm{pyr}^d(\{\mathbf{0}\})$.

The **product** of the polytopes $P \subset \mathbb{R}^d$ and $P' \subset \mathbb{R}^{d'}$ is

$$P \times P' := \{(\mathbf{p}, \mathbf{p}') \in \mathbb{R}^{d+d'} : \mathbf{p} \in P, \mathbf{p}' \in P'\}.$$

The new set $P \times P'$ is again a polytope.

Prism: The prism over P, denoted as $\mathrm{prism}(P)$, is the product of P with a line segment such as the simplex Δ_1. What we usually call a prism is the prism over a triangle. Just as a d-simplex can be obtained by taking d successive pyramids over a point, a d-cube can be constructed by taking d successive prisms over a point.

Bipyramid: This is the convex hull of P and any line segment that pierces the interior of P at precisely one point. To obtain an example, assume that the origin in $\mathbb{R}^d$ is in the interior of P and embed P in $\mathbb{R}^{d+1}$ as for the pyramid. Let $I := [-\mathbf{e}_{d+1}, \mathbf{e}_{d+1}] \subset \mathbb{R}^{d+1}$. Then $\mathrm{conv}(P, I)$ is an example of $\mathrm{bipyr}(P)$. Note that $C_d^\Delta = \mathrm{bipyr}(C_{d-1}^\Delta) = \cdots = \mathrm{bipyr}^d(\{0\})$.

Exercise 2.25. Let $P = \mathrm{conv}(\{\mathbf{v}_1, \ldots, \mathbf{v}_t\})$. Argue that for each of the above constructions it suffices to consider just $\mathbf{v}_1, \ldots, \mathbf{v}_t$. For example, argue that $\mathrm{pyr}(P) = \mathrm{conv}(\{\mathbf{v}_1, \ldots, \mathbf{v}_t, \mathbf{e}_{d+1}\})$.

Minkowski Sum: Suppose $d' \leq d$ and that P' is a d'-dimensional polytope in $\mathbb{R}^d$. The Minkowski sum of P and P' is defined to be $P + P' := \{\mathbf{p} + \mathbf{p}' : \mathbf{p} \in P, \mathbf{p}' \in P'\}$. This is again a polytope in $\mathbb{R}^d$.

Exercise 2.26. Let $L = [(0,0), (1,1)] \subset \mathbb{R}^2$, let C_2 be the square in $\mathbb{R}^2$ as before, and let Δ_2 be defined as $\mathrm{conv}\{(0,0), (1,0), (0,1)\}$. Confirm that $C_2 = [(0,0), (1,0)] + [(0,0), (0,1)]$. Draw $C_2 + \Delta_2$. Next draw the Minkowski sums $\Delta_2 + L$ and $C_2 + L$. You should get a 5-gon and 6-gon here. Argue that if $n \geq 3$ is odd, then an n-gon can be constructed as the Minkowski sum of a triangle and a collection of lines. If $n \geq 4$ is even, show that an n-gon can be constructed as the Minkowski sum of a collection of lines.

Minkowski sums of finitely many line segments are called *zonotopes*. Note that unlike prisms, pyramids and bipyramids, Minkowski sums of polytopes depend on the embedding of the summands.

There are many other basic constructions for building new polytopes from old ones. We will see some of them in the forthcoming chapters. For others we refer the reader to [**Grü03**] and [**Zie95**].

Chapter 3

Faces of Polytopes

In this chapter we continue with the basics of polytope theory. These facts can be found in books such as [**Grü03**] and [**Zie95**]. A short summary of polytope basics can be found in [**HRGZ97**].

Given any vector $\mathbf{c} \in \mathbb{R}^d$, we obtain a linear function $\phi_{\mathbf{c}} : \mathbb{R}^d \to \mathbb{R}$ such that $\mathbf{x} \mapsto \mathbf{c} \cdot \mathbf{x}$ where $\mathbf{c} \cdot \mathbf{x}$ is the dot product of $\mathbf{c}$ and $\mathbf{x}$. In optimization, one is often interested in the max (or min) value of this function over a specified polytope $P = \{\mathbf{x} \in \mathbb{R}^d : A\mathbf{x} \leq \mathbf{b}\} \subset \mathbb{R}^d$, leading to the *linear program*:

$$(3.1) \quad \max \{\mathbf{c} \cdot \mathbf{x} : \mathbf{x} \in P\} = \max \{\mathbf{c} \cdot \mathbf{x} : A\mathbf{x} \leq \mathbf{b}, \ \mathbf{x} \in \mathbb{R}^d\}.$$

If $\mathbf{c} = \mathbf{0}$, then the max value is simply 0. Otherwise, we can solve the above linear program geometrically as follows. Take the hyperplane $\{\mathbf{x} \in \mathbb{R}^d : \mathbf{c} \cdot \mathbf{x} = \beta\}$ for some value of β and then move it, in a perpendicular direction so that the translated version is parallel to the original, across P in the direction in which β increases. At a given position of this moving hyperplane (say with right-hand side value β'), all the points $\mathbf{x} \in \mathbb{R}^d$ that lie on the hyperplane have $\phi_{\mathbf{c}}(\mathbf{x}) = \mathbf{c} \cdot \mathbf{x} = \beta'$. Thus geometrically, the optimal (in this case, max) value of $\mathbf{c} \cdot \mathbf{x}$ over P is achieved by all points in P hit by the hyperplane immediately before the hyperplane goes past P. These points are the **optima** of the above linear program. We would like to understand this set.

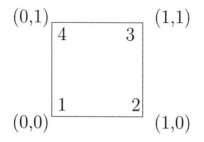

Figure 1. The unit square C_2.

Example 3.1. Take $P = C_2$ with its corners labeled as in Figure 1. Note that the linear functional $\phi_{(2,1)}(x_1, x_2) = 2x_1 + x_2$ is maximized at the corner $(1,1)$ of the square and nowhere else. See Figure 2(a). How much can we swing $\mathbf{c}$ on either side of $(2,1)$ keeping its base fixed and still have $c_1 x_1 + c_2 x_2$ be optimized at $(1,1)$? Note that we can go right as far as $(1,0)$ and left as far as $(0,1)$ and still have $(1,1)$ be the optimum of the linear program (3.1). The functional $\phi_{(1,0)}(\mathbf{x}) = x_1$ is in fact maximized by the whole right side of C_2 or in other words by the edge labeled 23. When we swing $\mathbf{c}$ slightly past $(1,0)$ to the right, we see that the unique optimum now is the corner $(1,0)$ labeled 2. Going all the way around with $\mathbf{c}$, we see that the optima of (3.1) move around the boundary of C_2 jumping from one vertex to a neighboring one at certain critical values of $\mathbf{c}$ that depend on the shape of C_2. In fact all of $\mathbb{R}^2$ breaks up as in Figure 2(d) depending on the optima of (3.1) over C_2.

For $\mathbf{c} \in \mathbb{R}^d$, let $m_{\mathbf{c}}(P) \in \mathbb{R}$ be the max value of the linear functional $\phi_{\mathbf{c}}(\mathbf{x})$ over P. The number $m_{\mathbf{c}}(P)$ is called the **optimal value** of the linear program (3.1).

Definition 3.2. For $\mathbf{c} \in \mathbb{R}^d \backslash \{0\}$, the hyperplane $H_{\mathbf{c}}(P) := \{\mathbf{x} \in \mathbb{R}^d : \mathbf{c} \cdot \mathbf{x} = m_{\mathbf{c}}(P)\}$ is called a **supporting hyperplane** of P, and for each $m \geq m_{\mathbf{c}}(P)$, the inequality $\mathbf{c} \cdot \mathbf{x} \leq m$ is said to be **valid** for P since all $\mathbf{x} \in P$ satisfy this inequality.

Example 3.3. The lines $H_{(2,1)}(C_2) = \{(x_1, x_2) : 2x_1 + x_2 = 3\}$ and $H_{(1,0)}(C_2) = \{(x_1, x_2) : x_1 = 1\}$ are supporting hyperplanes of C_2.

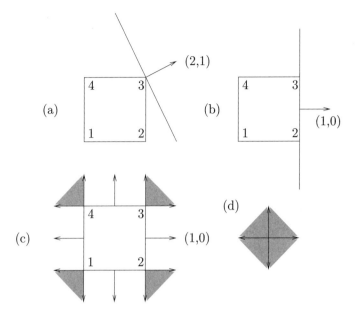

Figure 2. Maximizing linear functionals over C_2.

See Figures 2(a) and (b). The inequality $x_1 \le 2$ is valid for C_2 but does not provide a supporting hyperplane for C_2.

Definition 3.4. The intersection of a polytope $P \subset \mathbb{R}^d$ with a supporting hyperplane of P is called a **face** of P. More precisely, the face of P in direction $\mathbf{c} \ne \mathbf{0}$ is the intersection $\mathrm{face}_{\mathbf{c}}(P) := H_{\mathbf{c}}(P) \cap P$. These are the *non-trivial* faces of P.

When $\mathbf{c} = \mathbf{0}$, $H_{\mathbf{0}}(P) = \{\mathbf{x} \in \mathbb{R}^d : 0x_1 + \cdots + 0x_d = 0\} = \mathbb{R}^d$ and $\mathrm{face}_{\mathbf{0}}(P) = P \cap H_{\mathbf{0}}(P) = P$. Note that $H_{\mathbf{0}}(P)$ is not a hyperplane in $\mathbb{R}^d$. The empty set is always considered to be a face of P as well. The faces P and $\emptyset$ are the *trivial* faces of P.

Writing P in its $\mathcal{H}$-representation $P = \{\mathbf{x} \in \mathbb{R}^d : A\mathbf{x} \le \mathbf{b}\}$, we see that

$$
\begin{aligned}
\mathrm{face}_{\mathbf{c}}(P) =\ & \{\mathbf{x} \in \mathbb{R}^d : A\mathbf{x} \le \mathbf{b},\ \mathbf{c} \cdot \mathbf{x} = m_{\mathbf{c}}(P)\} \\
=\ & \{\mathbf{x} \in \mathbb{R}^d : A\mathbf{x} \le \mathbf{b},\ \mathbf{c} \cdot \mathbf{x} \le m_{\mathbf{c}}(P),\ -\mathbf{c} \cdot \mathbf{x} \le -m_{\mathbf{c}}(P)\}
\end{aligned}
$$

is again a polytope.

Definition 3.5.   (1) The **dimension** of a face of P is the dimension of the face as a polytope. The k-dimensional faces of P are called its k-faces. The 0-faces are called **vertices**, the 1-faces are called **edges** and the $(\dim(P) - 1)$-faces are called **facets**. The empty face of P is defined to have dimension -1.

  (2) The number of k-faces of P is denoted as $f_k(P)$ and is known as the kth **face number** (kth f-number) of P.

  (3) The **face vector** (f-vector) of P is the vector
$$f(P) := (f_0(P), f_1(P), \ldots, f_{\dim(P)}(P)).$$

Example 3.6. The vertices of C_2 are $(0,0), (1,0), (1,1)$ and $(0,1)$. The polytope $\mathrm{conv}(\{(0,0),(1,0)\})$ is an edge of C_2. This edge is also a facet of C_2. The unique 2-face of C_2 is C_2 itself. The face vector of C_2 is $f(C_2) = (4, 4, 1)$.

Note that the points in $\mathrm{face_c}(P)$ are precisely the optimal solutions to the linear program (3.1). Hence we call $\mathrm{face_c}(P)$ the **optimal face** of P in direction $\mathbf{c}$.

Recall that the $\mathcal{H}$-representation of $\mathrm{face_c}(P)$ showed that it is a polytope. By the main theorem of polytopes, we then know that it also has a $\mathcal{V}$-representation. In fact, $\mathrm{face_c}(P)$ is the convex hull of the vertices of P contained in it. Check this for C_2. This means that if we label the vertices of P by $1, 2, \ldots$, then each face of P can be labeled by the set of indices of the vertices of P whose convex hull is this face. For example, in Figure 1, the $\mathrm{face}_{(1,0)}(C_2)$ can be labeled by the set $\{2, 3\}$ or more compactly as 23. Since every face of a polytope P is the convex hull of a subset of the vertices of P, P has only finitely many faces in each dimension and hence $f_k(P)$ is a finite number for each $0 \leq k \leq \dim(P)$. Let us denote by $\mathcal{F}(P)$ the set of all faces of P. We can now partially order $\mathcal{F}(P)$ by the usual containment relation, $\subseteq$, on sets, creating a *partially ordered set* (**poset**) $(\mathcal{F}(P), \subseteq)$.

Definition 3.7. The partially ordered set $(\mathcal{F}(P), \subseteq)$ is called the **face lattice** of P.

Note that $\emptyset$ is the unique minimal element of $(\mathcal{F}(P), \subseteq)$ while P is the unique maximal element. Lattices are posets with the property

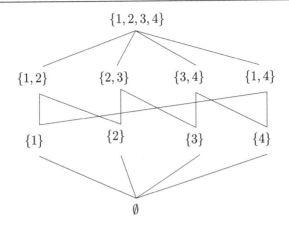

Figure 3. Face lattice of C_2.

that every pair of elements has a unique greatest lower bound and a unique lowest upper bound in the partial order. In particular, lattices have a unique minimal element and a unique maximal element. The face poset of a polytope is always a lattice. In Chapter 4 (see Figure 4 there) we will see posets that are not lattices.

Example 3.8. The face lattice of C_2 is shown in Figure 3. Such pictures of partially ordered sets are called *Hasse diagrams*.

By the combinatorics of a polytope one usually means the information about the polytope that is carried by its face lattice without concerning oneself with the actual geometric embedding of the polytope. For instance note that C_2 has the same face lattice as any square in $\mathbb{R}^2$ or for that matter any square in $\mathbb{R}^d$. In fact, it has the same face lattice as any (planar) quadrilateral in $\mathbb{R}^d$. Thus combinatorially all quadrilaterals are equivalent to C_2 although they may vary wildly in actual embedding into some $\mathbb{R}^d$. The face lattice is an abstraction of the polytope and by its nature forgets a lot of information about the polytope it came from. However, it does retain essential "combinatorial" information such as how many vertices the polytope had, which vertices define which face and which faces are contained in which other faces. This is called the *face incidence* information.

Definition 3.9. Two polytopes P and Q are **combinatorially isomorphic** or **combinatorially equivalent** if $\mathcal{F}(P)$ and $\mathcal{F}(Q)$ are isomorphic as partially ordered sets.

We have not defined what it means to be isomorphic as partially ordered sets. We can just take that to mean that both $\mathcal{F}(P)$ and $\mathcal{F}(Q)$ have the same Hasse diagram up to relabeling of the vertices.

Exercise 3.10. Draw the face lattices of the 3-cube C_3 and the octahedron C_3^Δ. Do you see a relationship between the two?

Exercise 3.11. Let $t \in \mathbb{N}$. Recall that the power set of $[t] = \{1, 2, \ldots, t\}$, denoted as $2^{[t]}$, is the set of all subsets of $[t]$. This power set is partially ordered by $\subseteq$.

(1) Is there a polytope whose face lattice is $(2^{[t]}, \subseteq)$?

(2) Does every partially ordered subset of $(2^{[t]}, \subseteq)$ appear as the face lattice of a polytope? If not, do you see conditions that are necessary for a subset of $2^{[t]}$ to be the face lattice of a polytope?

Definition 3.12. A d-dimensional polytope P is **simple** if every vertex of P is incident to d edges of P or, equivalently, if every vertex of P lies on precisely d facets of P.

Definition 3.13. A d-dimensional polytope P is **simplicial** if for $0 \leq k \leq \dim(P) - 1$, each k-face of P is combinatorially isomorphic to a k-simplex.

Note that a simplicial polytope does not have to be a simplex since the condition on k-faces is only required to hold for $k \leq \dim(P) - 1$.

Example 3.14. The cubes C_d are simple polytopes while the cross-polytopes C_d^Δ are simplicial polytopes.

Exercise 3.15. Can you tell from the face lattice of a polytope whether the polytope is simple or simplicial?

Exercise 3.16. (1) Show that a polygon (a 2-polytope) is both simple and simplicial.

(2) Construct a polytope that is neither simple nor simplicial.

Definition 3.17. If $P \subset \mathbb{R}^d$ is a d-polytope with the origin in its interior, then the **polar** of P is the d-polytope

$$P^\Delta := \{\mathbf{y} \in \mathbb{R}^d : \mathbf{y} \cdot \mathbf{x} \leq 1 \text{ for all } \mathbf{x} \in P\}.$$

Check that cubes are polar to cross-polytopes. If you use the definition of C_d^Δ that was given in the last chapter, what exactly is the $\mathcal{H}$-representation of the cube that is polar to C_d^Δ?

There are many nice relationships between a polytope P and its polar P^Δ. We list some of them below.

(1) If $P = \mathrm{conv}(\{\mathbf{v}_1, \ldots, \mathbf{v}_t\}) \subset \mathbb{R}^d$, then $P^\Delta = \{\mathbf{x} \in \mathbb{R}^d : \mathbf{v}_i \cdot \mathbf{x} \leq 1, \text{ for all } i = 1, \ldots, t\}$.

(2) $(P^\Delta)^\Delta = P$.

(3) The polars of simple polytopes are simplicial and the polars of simplicial polytopes are simple.

(4) The face lattices $\mathcal{F}(P)$ and $\mathcal{F}(P^\Delta)$ are anti-isomorphic. This means that there is a bijection between the k-faces of P and the $(d - k - 1)$-faces of P^Δ, where we assume that $\dim(P) = \dim(P^\Delta) = d$, that also "inverts" the inclusions. Informally, the Hasse diagram of $(\mathcal{F}(P), \subseteq)$ can be gotten by rotating the Hasse diagram of $(\mathcal{F}(P^\Delta), \subseteq)$ by 180 degrees.

In the rest of this chapter, we focus on a special class of polytopes called *cyclic polytopes*. The cyclic polytope $C_d(n)$ is a simplicial d-polytope with n vertices. The most well-known property of the cyclic polytope is that it provides an upper bound on face numbers of simplicial polytopes with the same dimension and number of vertices.

Theorem 3.18. (*The Upper Bound Theorem*) *Let P be any simplicial d-polytope with n vertices. Then $f_i(P) \leq f_i(C_d(n))$ for every $0 \leq i \leq d - 1$.*

The construction of $C_d(n)$ is possible using techniques from linear algebra and its most satisfying combinatorial property is a simple description of its facets. We follow the treatment in [**Zie95**].

The *moment curve* in $\mathbb{R}^d$ is defined by the function $\phi : \mathbb{R} \to \mathbb{R}^d$, $t \mapsto (t, t^2, \ldots, t^d)$. If we fix d, n and $t_1 < t_2 < \ldots < t_n \in \mathbb{R}$, we can define the d-dimensional **cyclic polytope** with n vertices as

$$C_d(n) := \mathrm{conv}(\{\phi(t_1), \phi(t_2), \ldots, \phi(t_n)\}).$$

We say *the* cyclic polytope since any such choice of t_i's will yield the same face lattice. We will show the following.

(1) $C_d(n)$ is a simplicial polytope.

(2) The facets of $C_d(n)$ satisfy *Gale's evenness condition*: $S = \{i_1, i_2, \ldots, i_d\} \subset [n]$ indexes a facet of $C_d(n)$ if and only if for all $i, j \notin S$ and $i < j$,

$$2 \text{ divides } |\{k : k \in S, i < k < j\}|.$$

(3) Every $\mathcal{I} \subset [n]$ with $|\mathcal{I}| \leq \lfloor \frac{d}{2} \rfloor$ indexes a face of $C_d(n)$. More commonly, $C_d(n)$ is said to be $\lfloor \frac{d}{2} \rfloor$-*neighborly*.

Definition 3.19. A polytope P is k-**neighborly** if any set of k or fewer vertices of P is the vertex set of a face of P.

Example 3.20. Using PORTA [**CL**], we can compute the facets of the cyclic polytope $C_4(7)$ in dimension four with seven vertices. The input file consists of the seven vertices listed under CONV_SECTION, as shown below.

```
DIM = 4

CONV_SECTION
1 1 1 1
2 4 8 16
3 9 27 81
4 16 64 256
5 25 125 625
6 36 216 1296
7 49 343 2401
END
```

The output file has two parts. The facet inequalities of $C_4(7)$ are listed at the top under INEQUALITIES_SECTION. After that, we get the *strong validity table* of $C_4(7)$, which is a table whose columns index the seven vertices and whose rows index the facets of $C_4(7)$. The stars in a row indicate which vertices of $C_4(7)$ lie on the facet indexing that row.

```
DIM = 4

VALID
7 49 343 2401
```

```
INEQUALITIES_SECTION
(   1) -317x1+125x2-19x3+x4 <= -210
(   2) -223x1+ 99x2-17x3+x4 <= -140
(   3) -145x1+ 75x2-15x3+x4 <=  -84
(   4) - 83x1+ 53x2-13x3+x4 <=  -42
(   5) + 50x1- 35x2+10x3-x4 <=   24
(   6) + 78x1- 49x2+12x3-x4 <=   40
(   7) +112x1- 65x2+14x3-x4 <=   60
(   8) +152x1- 83x2+16x3-x4 <=   84
(   9) +154x1- 71x2+14x3-x4 <=  120
(  10) +216x1- 91x2+16x3-x4 <=  180
(  11) +288x1-113x2+18x3-x4 <=  252
(  12) +342x1-119x2+18x3-x4 <=  360
(  13) +450x1-145x2+20x3-x4 <=  504
(  14) +638x1-179x2+22x3-x4 <=  840
END
```

```
strong validity table :
\ P       |           |
 \ O      |           |
I \ I     |           |
 N \ N    | 1      6  | #
  E \ T   |           |
   Q \ S  |           |
    S \   |           |
     \ |             |
-----------------------
1        | *...* ** :  4
2        | *..** .* :  4
3        | *.**. .* :  4
4        | ***.. .* :  4
5        | ****. .. :  4
6        | **.** .. :  4
7        | **..* *. :  4
8        | **... ** :  4
9        | .**** .. :  4
```

```
10        | .**.* *. :    4
11        | .**.. ** :    4
12        | ..*** *. :    4
13        | ..**. ** :    4
14        | ...** ** :    4

          . . . . . . . . . . . . . .
#         | 88888 88
```

There are fourteen facets in all, the first of which contains the vertices $1, 5, 6, 7$. Verify that properties (1), (2) and (3) are satisfied.

Exercise 3.21. Let $r_0, r_1, \cdots, r_d \in \mathbb{R}$.

    (a) Prove Vandermonde's identity:

$$\det \begin{pmatrix} 1 & 1 & \cdots & 1 \\ \phi(r_0) & \phi(r_1) & \cdots & \phi(r_d) \end{pmatrix} = \prod_{0 \leq i < j \leq d} (r_j - r_i).$$

    (b) Deduce that no $d+1$ points on the moment curve are affinely dependent.

    (c) Conclude that $C_d(n)$ must be a simplicial polytope.

Exercise 3.22. Let $S = \{i_1, i_2, \ldots, i_d\} \subset [n]$.

    (a) Write down the equation $H_S(\mathbf{x}) = m$ that defines the hyperplane spanned by $\phi(t_{i_1}), \phi(t_{i_2}), \ldots, \phi(t_{i_d})$.

    (b) Suppose S indexes a facet of $C_d(n)$. What relationship can you find between the hyperplane $H_S(\mathbf{x}) = m$ and $\{\phi(t_j) : j \notin S\}$?

    (c) If S indexes a facet, draw a picture of how $H_S(\mathbf{x}) = m$ intersects the moment curve $\phi(t)$. Begin by drawing this for $d = 2$.

Exercise 3.23. Deduce that $C_d(n)$ is $\lfloor \frac{d}{2} \rfloor$-neighborly. **Hint:** Pick any $I \subset [n]$ with $|I| = \lfloor \frac{d}{2} \rfloor$. Using Gale's evenness condition, can you find a facet that contains I?

Chapter 4

Schlegel Diagrams

We usually draw a polytope on paper by drawing its vertices and edges. For instance, the square C_2 was drawn this way in the last chapter. This is called the 1-skeleton of the polytope (all faces of dimension at most one). For the combinatorics of the polytope it is enough to think of this 1-skeleton as an abstract graph $G = (V, E)$ where V is the vertex set of the graph G and E is the edge set of G. Even though a graph may have many different drawings, its combinatorics is fixed and thus it carries important information.

Definition 4.1. The **graph of a polytope** P is the abstract graph $G(P) = (V(P), E(P))$, where $V(P) = \{v_1, \ldots, v_t\}$ is a set of labels for the vertices of P (v_i is a label for the vertex $\mathbf{v}_i$ of P). The elements of $E(P)$ are the sets $\{v_i, v_j\}$ where $\mathrm{conv}(\{\mathbf{v}_i, \mathbf{v}_j\})$ is an edge of P.

Example 4.2. Figure 1 shows two different drawings of the graph of a 3-cube.

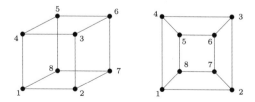

Figure 1. Two drawings of the graph of a 3-cube.

Definition 4.3. A graph $G = (V, E)$ is said to be **planar** if it can be drawn in the plane in such a way that no two edges meet except possibly at their end points.

Clearly, the graph of every two-dimensional polytope is planar. Figure 1 shows that the graph of a 3-cube is planar. How about other three-dimensional polytopes? The following famous theorem from about 100 years ago answers this question.

Theorem 4.4. Steinitz Theorem. *G is the graph of a three-dimensional polytope if and only if it is simple, planar and 3-connected.*

Definition 4.5.   (1) A graph G is **simple** if it has no loops and no multiple edges. A **loop** is an edge of the form $\{v, v\}$, where v is the label of a vertex of the graph.

    (2) A simple graph G is k-**connected** if it remains connected after the removal of any $k-1$ or fewer vertices. (All the edges incident to these vertices are also removed.) Equivalently, a graph G is k-connected if there are k edge disjoint paths between any two vertices of G.

Exercise 4.6. Draw a planar embedding of the graph of an octahedron. How about an icosahedron or a dodecahedron?

We next state three well-known facts and conjectures about graphs of polytopes that the reader might find interesting.

Theorem 4.7. *(Balinski) If P is a d-dimensional polytope, then $G(P)$ is d-connected.*

The following is a weaker result.

Exercise 4.8. Let P be a d-dimensional polytope. Argue that every vertex in $G(P)$ has degree at least d. Give an example to show that this is not strong enough to show d-connectedness of $G(P)$.

Theorem 4.9. *(Blind and Mani) If P and Q are simple polytopes with their graphs $G(P)$ and $G(Q)$ being (combinatorially) isomorphic, then P and Q are combinatorially isomorphic as polytopes.*

Conjecture 4.10. Hirsch conjecture. *Let P be a d-dimensional polytope with n facets. If u and v are any two vertices in $G(P)$, then* *there exists a path in $G(P)$ going from u to v that contains at most $n - d$ edges.*

Exercise 4.11. Verify the Hirsch conjecture for the 3-cube, 4-cube and any other polytope that takes your fancy.

The Steinitz theorem is a very satisfactory understanding of the graphs of three-dimensional polytopes. In fact, for every planar 3-connected simple graph G there is only one 3-polytope P (up to combinatorial isomorphism) with $G(P) = G$. Moreover, this P can be embedded in $\mathbb{R}^3$ with integer coordinates. This may seem somewhat trivial right now, but in Chapter 6 we will see that it is not always possible to find integer coordinates for the vertices of a polytope whose combinatorics (face lattice) has been specified.

Can we extend our graph technique to visualize four-dimensional polytopes? To do this, let's rethink how we drew the planar graph of the 3-cube in Figure 1. What we did can be expressed roughly as follows. Imagine you are an ant looking into a hollow 3-cube through a tiny pinhole in the center of one square facet of the cube. This facet is your 360^0 horizon and we draw it first. In the figure, this is the facet labeled 1234. As you look into the cube, you see the opposite square facet to 1234 as a small square in the distance. Drawing what we see, we put this small facet 5678 in the center of the square 1234. Now you see the edges that connect the front facet to the back facet as the four segments $45, 36, 18$ and 27. These edges are actually parallel, but due to their large lengths relative to you, you see them as parts of four lines that will eventually meet at a point at infinity. The resulting figure is a drawing of the graph of the polytope through a facet of the polytope. Yet another way of thinking about this is that you can imagine that the 3-cube is made of rubber. Puncture one facet in the middle and then stretching this hole wide open, flatten the cube onto your paper. The edges of this flattened polytope will provide the graph drawing.

Can we do this for 4-polytopes as well? We want to first draw one facet (which is now a three-dimensional polytope) and then draw the rest of the polytope inside this facet. Let's try it on the following 4-simplex.

Example 4.12. Consider the 4-simplex

$$P = \operatorname{conv}(\{(0,0,0,0),(1,0,0,0),(0,1,0,0),(0,0,1,0),(0,0,0,1)\}).$$

Its $\mathcal{H}$-representation is

$$P = \left\{ (x_1, x_2, x_3, x_4) \in \mathbb{R}^4 : \begin{array}{c} x_1 \geq 0 \\ x_2 \geq 0 \\ x_3 \geq 0 \\ x_4 \geq 0 \\ x_1 + x_2 + x_3 + x_4 \leq 1 \end{array} \right\}.$$

The five inequalities given above all define facets of P. By plugging in the vertices of P into the five inequalities, we can see which vertices lie on which facets. This gives the following **facet-vertex incidence table** for P. This computation was done using PORTA although we could have also done it by hand.

```
strong validity table :
\ P       |       |
 \ O      |       |
I \ I     |       |
 N \ N    | 1     | #
  E \ T   |       |
   Q \ S  |       |
    S \   |       |
     \ |        |
--------------------
1         | *.*** :   4
2         | **.** :   4
3         | ***.* :   4
4         | ****. :   4
5         | .**** :   4

           . . . . . . . . . . .
#         | 44444
```

We see immediately that each facet is a 3-simplex (since it is the convex hull of four points all of which are vertices). This implies that any two vertices form an edge of P since every pair of vertices appears as vertices of some facet which is a simplex, and also every three vertices of P form a triangular 2-face of P as well. Can you see all of this in the above table?

Let's start by first drawing the facet 2345 (in solid lines) and then the rest of the polytope inside this facet (with dashed lines). The vertex 1 is placed inside the tetrahedron 2345 and then we draw all the tetrahedral facets of the simplex involving 1 with dashed lines. This gives Figure 2. Check that there are four three-dimensional simplices inside the outer tetrahedron which are the perspective drawings of the four facets of the 4-simplex that we can see through the facet 2345.

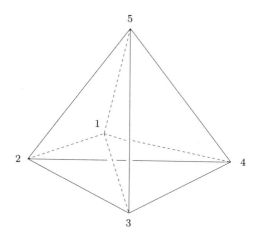

Figure 2. A 4-simplex.

Exercise 4.13. (1) Can the graph of a 4-polytope be planar?

(2) Can you draw the 4-cube as we drew the 4-simplex?

Diagrams such as the above are known as *Schlegel diagrams* and are extremely useful for visualizing 4-polytopes. We define them formally now. See [**Zie95**, Lecture 5] for many pictures and much of the material below.

Definition 4.14. A **polyhedral complex** $\mathcal{C}$ is a finite collection of polyhedra in $\mathbb{R}^d$ such that

(1) the empty polyhedron is in $\mathcal{C}$,

(2) if $P \in \mathcal{C}$, then all faces of P are also in $\mathcal{C}$,

(3) if P and Q are in $\mathcal{C}$, then $P \cap Q$ is a face of both P and Q.

Example 4.15. In Figure 3, the first is a polyhedral complex and the second is not. Why?

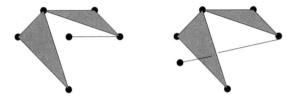

Figure 3. A polyhedral complex: example and non-example.

The dimension of a polyhedral complex $\mathcal{C}$ is the maximum dimension of a polyhedron in $\mathcal{C}$. The **support** of $\mathcal{C}$ is the set $|\mathcal{C}| = \bigcup \{P : P \in \mathcal{C}\}$. Just as we drew face lattices of polytopes, a polyhedral complex $\mathcal{C}$ has a **face poset** which is the set of all the polyhedra in $\mathcal{C}$ ordered by $\subseteq$. The face poset of the polyhedral complex on the left in Figure 3 is shown in Figure 4. Note that this poset is not a lattice.

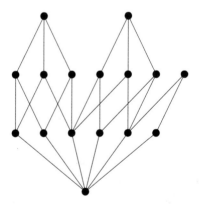

Figure 4. Face poset of the polyhedral complex in Figure 3.

Exercise 4.16. Can you assign face labels to the poset in Figure 4 to verify that it is the face poset of the polyhedral complex in Figure 3?

We are interested in polyhedral complexes since a polytope P gives rise to some natural polyhedral complexes which play an important role in the study of polytopes.

Definition 4.17. Let P be a polytope.

(1) The **complex** $\mathcal{C}(P)$ **of the polytope** P is the polyhedral complex of all faces of P. The face poset of $\mathcal{C}(P)$ is the face lattice of P.

(2) The **boundary complex** $\partial(\mathcal{C}(P))$ is the polyhedral complex of all the proper faces of P along with the empty face.

(3) A **polytopal subdivision** of P is a polyhedral complex $\mathcal{C}$ with support P in which all the polyhedra are polytopes. If all the maximal polytopes in the subdivision are simplices, the subdivision is called a **triangulation** of P.

For instance Figure 5 shows two polytopal subdivisions of a pentagon. The second subdivision is a triangulation. Notice that new vertices might be introduced when we subdivide a polytope.

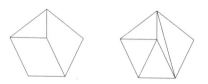

Figure 5. Polytopal subdivisions of a pentagon.

Suppose F is a facet of the full-dimensional polytope $P = \{\mathbf{x} \in \mathbb{R}^d : A\mathbf{x} \leq \mathbf{b}\}$ and $\mathrm{aff}(F) := \{\mathbf{x} \in \mathbb{R}^d : \mathbf{a} \cdot \mathbf{x} = b\}$ is the affine hull of F. Note that $\mathrm{aff}(F)$ is a hyperplane in $\mathbb{R}^d$. Assume that for each facet G of P, P is contained in the halfspace $\mathrm{aff}(G)^-$. We say that a point $\mathbf{y}$ is **beyond** the facet F if $\mathbf{y} \in \mathrm{aff}(F)^+$, $\mathbf{y} \notin P$, but $\mathbf{y}$ is in the interior of the halfspace $\mathrm{aff}(G)^-$ for all other facets G of P.

Definition 4.18. ([**Zie95**, Definition 5.5]) Let P be a d-polytope in $\mathbb{R}^d$ and let F be a facet of P defined by the inequality $\mathbf{a} \cdot \mathbf{x} \leq b$. Then $\mathrm{aff}(F) = \{\mathbf{x} \in \mathbb{R}^d : \mathbf{a} \cdot \mathbf{x} = b\}$ is the hyperplane spanned by F. Choose a point $\mathbf{y}_F$ beyond F. For $\mathbf{x} \in P$ define the function $p(\mathbf{x})$ as follows (see Figure 6):

$$p(\mathbf{x}) := \mathbf{y}_F + \frac{b - \mathbf{a}\mathbf{y}_F}{\mathbf{a}\mathbf{x} - \mathbf{a}\mathbf{y}_F}(\mathbf{x} - \mathbf{y}_F).$$

The **Schlegel diagram** of P based at the facet F, denoted as $\mathcal{C}(P, F)$, is the image under p of all proper faces of P other than F.

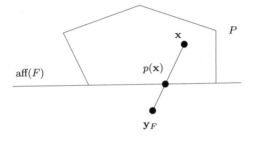

Figure 6. Definition of $p(\mathbf{x})$.

Proposition 4.19. ([**Zie95**, Proposition 5.6]) *The Schlegel diagram of P based at the facet F is a polytopal subdivision of F that is combinatorially equivalent to the complex $\mathcal{C}(\partial(P)\backslash\{F\})$ of all proper faces of P other than F.*

Proof. For a face G of P different from F, the set

$$C_G := \{\mathbf{y}_F + \lambda(\mathbf{x} - \mathbf{y}_F) : \mathbf{x} \in G, \lambda \geq 0\}$$

is a *cone* with vertex $\mathbf{y}_F$. See Figure 7. If G is a proper face of P, then it is contained in the hyperplane $\mathrm{aff}(G) = \{\mathbf{x} \in \mathbb{R}^d : \mathbf{a}' \cdot \mathbf{x} = b'\}$ that does not contain $\mathbf{y}_F$. Thus the face lattice of G is isomorphic to the face lattice of C_G. The intersection of C_G with $\mathrm{aff}(F)$ also has face lattice isomorphic to C_G and hence to G. However this intersection is $p(G)$. Thus the face lattice of G is isomorphic to the face lattice of $p(G)$. $\qquad\square$

Example 4.20. Figure 1(b) is a Schlegel diagram of a 3-cube while Figure 2 is the Schlegel diagram of a 4-simplex.

If the Schlegel diagram of a polytope P based at the facet F is given as a polytopal complex $\mathcal{D}$, then the face lattice of P is the face

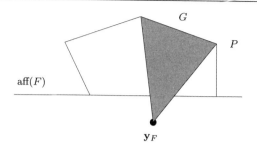

Figure 7. Definition of C_G.

poset of $\mathcal{D}$ along with the faces F and P added in. How does F sit
in this poset? For a face G in $\mathcal{D}$, $G \subseteq F$ if and only if G is a face of
F. The Schlegel diagram thus completely encodes the combinatorics
of a d-dimensional polytope into a $(d-1)$-dimensional complex. This
is especially useful for $d \leq 4$.

Exercise 4.21. What polytope has the Schlegel diagram shown in
Figure 8? Draw the face lattice of this polytope. Does this polytope
have a different Schlegel diagram if you look through a different facet?

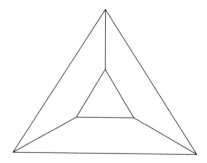

Figure 8. Figure for Exercise 4.21.

Now comes an unexpected subtlety. The polytopal complex shown
in Figure 9 is not the Schlegel diagram of any 3-polytope! If you look
at the diagram closely, you might convince yourself that as a graph,

it is the graph of an octahedron. It even looks like the Schlegel diagram of the octahedron seen through a triangular facet. However this figure lacks an important property of Schlegel diagrams that we now describe.

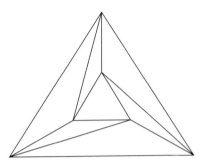

Figure 9. A non-Schlegel diagram.

Definition 4.22. ([**Zie95**, Definition 5.3]) A subdivision $\mathcal{C}$ of a polytope $Q \subseteq \mathbb{R}^d$ is **regular** if it arises from a polytope $P \subseteq \mathbb{R}^{d+1}$ in the following way:

(1) The polytope Q is the image $\pi(P) = Q$ of the polytope P via the projection

$$\pi : \mathbb{R}^{d+1} \to \mathbb{R}^d, \ \begin{pmatrix} \mathbf{x} \\ x_{d+1} \end{pmatrix} \mapsto \mathbf{x},$$

which deletes the last coordinate.

(2) The complex $\mathcal{C}$ is the projection under π of all the **lower faces** of P. We call F a lower face of P if for every $\mathbf{x} \in F$ and a $\lambda > 0$, $\mathbf{x} - \lambda \mathbf{e}_{d+1} \notin P$. Informally, they are the faces that you can see from Q if you "look up" at P from Q. See Figure 10.

Proposition 4.23. ([**Zie95**, Proposition 5.9]) *If $\mathcal{D}$ is a Schlegel diagram, then $\mathcal{D}$ is a regular subdivision of $|\mathcal{D}|$, the support of $\mathcal{D}$.*

Exercise 4.24. Can you see why the polytopal complex shown in Figure 9 is not a regular subdivision? You have to argue that the picture you see cannot be the projection of the lower faces of any 3-polytope.

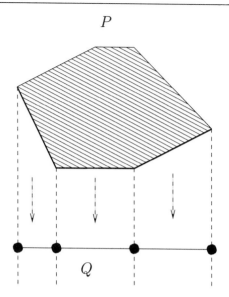

Figure 10. Lower faces of P (which are thickened) induce a subdivision of Q.

Exercise 4.25. Is every regular subdivision of a polytope the Schlegel diagram of some polytope?

Thus we see that we have to be a bit careful when dealing with Schlegel diagrams. Not everything that looks like a Schlegel diagram is indeed a Schlegel diagram.

Exercise 4.26. Draw the Schlegel diagrams of the cyclic polytopes $C_4(6)$ and $C_4(7)$.

Chapter 5

Gale Diagrams

In the last chapter we saw techniques for visualizing four-dimensional polytopes via their Schlegel diagrams. In this chapter, we will see that we can actually visualize even higher-dimensional polytopes as long as they do not have too many vertices. We do this via a tool called the Gale diagram of the polytope.

Consider n points $\mathbf{v}_1, \ldots, \mathbf{v}_n$ in $\mathbb{R}^{d-1}$ whose affine hull has dimension $d-1$ and the matrix

$$A := \begin{pmatrix} 1 & 1 & \cdots & 1 \\ \mathbf{v}_1 & \mathbf{v}_2 & \cdots & \mathbf{v}_n \end{pmatrix} \in \mathbb{R}^{d \times n}.$$

A basic fact of affine linear algebra is that the vectors $\mathbf{v}_1, \ldots, \mathbf{v}_n$ are affinely independent (see below) if and only if the vectors

$$(1, \mathbf{v}_1), \ldots, (1, \mathbf{v}_n)$$

are linearly independent. If the dimension of $\mathrm{aff}(\mathbf{v}_1, \ldots, \mathbf{v}_n)$ is $d-1$, then there are d affinely independent vectors in this collection, which in turn implies that the rank of A is d. Hence the dimension of the kernel of A is $n - d$. Recall that the kernel of A is the linear subspace

$$\ker_{\mathbb{R}}(A) := \{\mathbf{x} \in \mathbb{R}^n : A\mathbf{x} = \mathbf{0}\}.$$

Note that $\mathbf{x} \in \ker_{\mathbb{R}}(A)$ if and only if (1) $\sum_{i=1}^{n} \mathbf{v}_i x_i = \mathbf{0}$ and (2) $\sum_{i=1}^{n} x_i = 0$.

Definition 5.1.   (1) Any vector $\mathbf{x}$ with properties (1) and (2) is called an **affine dependence relation** on $\mathbf{v}_1, \ldots, \mathbf{v}_n$.

    (2) If $\mathbf{x}$ satisfies only (1), then it would be a **linear dependence relation** on $\mathbf{v}_1, \ldots, \mathbf{v}_n$.

    (3) If $\mathbf{x} = 0$ is the only solution to (1) and (2), then $\mathbf{v}_1, \ldots, \mathbf{v}_n$ are said to be **affinely independent**.

Let $B_1, \ldots, B_{n-d} \in \mathbb{R}^n$ be a basis for the vector space $\ker_{\mathbb{R}}(A)$. If we organize these vectors as the columns of an $n \times (n - d)$ matrix

$$B := \begin{pmatrix} B_1 & B_2 & \cdots & B_{n-d} \end{pmatrix},$$

we see that $AB = \mathbf{0}$.

Definition 5.2. Let $\mathcal{B} = \{\mathbf{b}_1, \ldots, \mathbf{b}_n\} \subset \mathbb{R}^{n-d}$ be the n ordered rows of B. Then $\mathcal{B}$ is called a **Gale transform** of $\{\mathbf{v}_1, \ldots, \mathbf{v}_n\}$. The associated **Gale diagram** of $\{\mathbf{v}_1, \ldots, \mathbf{v}_n\}$ is the *vector configuration* $\mathcal{B}$ drawn in $\mathbb{R}^{n-d}$.

Later, we will see a more general definition of Gale diagrams. Since the columns of B can be any basis of $\ker_{\mathbb{R}}(A)$, Gale transforms are not unique. However all choices of B differ by multiplication by a non-singular matrix and we will be happy to choose one basis of $\ker_{\mathbb{R}}(A)$ and call the resulting $\mathcal{B}$, *the* Gale transform of $\{\mathbf{v}_1, \ldots, \mathbf{v}_n\}$.

Example 5.3. Let $\{\mathbf{v}_i\}$ be the vertices of the triangular prism shown in Figure 1. Then

$$A = \begin{pmatrix} 1 & 1 & 1 & 1 & 1 & 1 \\ 0 & 1 & 0 & 0 & 1 & 0 \\ 0 & 0 & 1 & 0 & 0 & 1 \\ 0 & 0 & 0 & 1 & 1 & 1 \end{pmatrix}.$$

Computing a basis for the kernel of A, we get

$$B^t = \begin{pmatrix} 0 & 1 & -1 & 0 & -1 & 1 \\ 1 & 0 & -1 & -1 & 0 & 1 \end{pmatrix}$$

where B^t is the transpose of B. The Gale transform $\mathcal{B}$ is the vector configuration consisting of the columns of B^t (or the rows of B). In our example, $\mathcal{B} = \{\mathbf{b}_1 = (0, 1), \mathbf{b}_2 = (1, 0), \mathbf{b}_3 = (-1, -1), \mathbf{b}_4 = (0, -1), \mathbf{b}_5 = (-1, 0), \mathbf{b}_6 = (1, 1)\}$. The Gale diagram is shown in Figure 2.

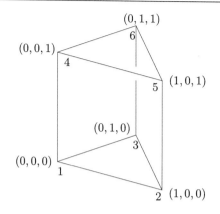

Figure 1. Triangular prism.

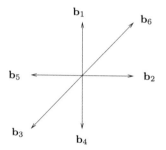

Figure 2. Gale diagram of the vertices of the triangular prism in Figure 1.

The labeling is very important in the construction of a Gale transform. We label column i of B^t as $\mathbf{b}_i$.

The main goal of this chapter will be to understand how to read off the face lattice of the $(d-1)$-polytope $P = \text{conv}(\{\mathbf{v}_1, \dots, \mathbf{v}_n\})$ from the Gale diagram of $\{\mathbf{v}_1, \dots, \mathbf{v}_n\}$. If $\{\mathbf{v}_j : j \in J\}$ are all the vertices on a face of P for some $J \subseteq [n]$, it is convenient to simply label this face by J. Here is a very important characterization of faces.

Lemma 5.4. *Let* $P = \text{conv}(\{\mathbf{v}_1, \dots, \mathbf{v}_n\})$. *Then* $J \subseteq [n]$ *is a face of* P *if and only if*

$$\text{conv}(\{\mathbf{v}_j : j \in [n] \setminus J\}) \cap \text{aff}(\{\mathbf{v}_j : j \in J\}) = \emptyset.$$

Let us illustrate this condition on an example first. In Figure 3(a), note that 15 is a face of the pentagon and that $\text{conv}(\{\mathbf{v}_2, \mathbf{v}_3, \mathbf{v}_4\})$ does not intersect the affine hull of the face 15. On the other hand, 14 is not a face of the pentagon and indeed $\text{conv}(\{\mathbf{v}_2, \mathbf{v}_3, \mathbf{v}_5\})$ does intersect the affine hull of the **non-face** 14. See Figure 3(b).

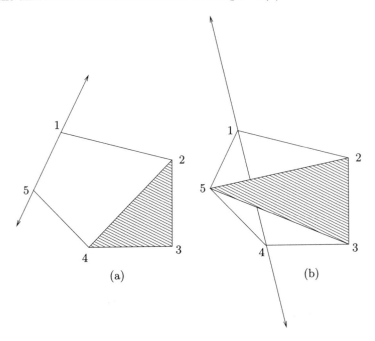

Figure 3. Condition in Lemma 5.4.

Proof. We may assume that P is a full-dimensional polytope. If J is a face of P, then by definition of a face, both J and $\text{aff}(J)$ lie on a supporting hyperplane H of P. Choose a supporting hyperplane H that contains J but does not contain any higher-dimensional face of P. One way to do this would be to let the normal vector of H be the sum of the normal vectors of the affine spans of the facets containing J. We may assume without loss of generality that P lies in the halfspace H^-. Since no $\mathbf{v}_j$, $j \notin J$, lies on the face $\text{conv}(\{\mathbf{v}_j : j \in J\})$ of P, $\text{conv}(\{\mathbf{v}_j : j \in [n]\backslash J\})$ lies in the interior of H^-, which proves one direction of the lemma. Conversely, if $\text{conv}(\{\mathbf{v}_j : j \in [n]\backslash J\}) \cap$

aff($\{\mathbf{v}_j : j \in J\}) = \emptyset$, then P lies in one halfspace defined by the hyperplane H obtained by extending aff($\{\mathbf{v}_j : j \in J\}$) which is thus a supporting hyperplane of P. This shows that J is a face of P. $\quad\square$

You might wonder why the lemma was not stated in the seemingly stronger form: $J \subseteq [n]$ is a face of P if and only if conv($\{\mathbf{v}_j : j \in [n]\backslash J\}) \cap$ conv($\{\mathbf{v}_j : j \in J\}) = \emptyset$. The above form of the lemma is what is needed to prove the main theorem below.

Definition 5.5. Call $[n]\backslash J$ a **co-face** of P if J is a face of P.

Note that a co-face is not the same as a non-face. In the triangular prism in Figure 1, 123 is both a face and a co-face. (The labeling of the vertices of the prism was fixed by how we ordered them to create the matrix A.)

In order to understand our main theorem, we need to define formally what we mean by the *interior* and *relative interior* of a polytope. The **interior** of a polytope in $\mathbb{R}^d$ is the set of all points in the polytope such that we can fit a d-dimensional ball centered at this point, of infinitesimal (as small as you wish but positive) radius, entirely inside the polytope. A polytope has an interior if and only if it is full-dimensional. For instance the interior of C_2 is the set

$$\text{int}(C_2) = \{(x_1, x_2) \in \mathbb{R}^2 : 0 < x_1 < 1, \ 0 < x_2 < 1\}.$$

The line segment conv($\{(1,0),(0,1)\}) \subset \mathbb{R}^2$ does not have an interior since there is no point on this segment such that a two-dimensional ball centered at this point will be contained in the line segment. However, this line segment does have an interior if we think of it as a polytope in its affine hull, where it is a full-dimensional polytope. This is known as the **relative interior** of the line segment. In our example, relint(conv($\{(1,0),(0,1)\}$)) equals

$$\{(x_1, x_2) \in \mathbb{R}^2 : x_1 + x_2 = 1, x_1 > 0, x_2 > 0\}.$$

We now come to the main theorem of this chapter. The proof of this theorem is taken from [**Grü03**, page 88].

Theorem 5.6. *Let* $P = \text{conv}(\{\mathbf{v}_1, \ldots, \mathbf{v}_n\})$, $\mathbf{v}_i \in \mathbb{R}^{d-1}$, *and let* $\mathcal{B}$ *be the Gale transform of* $\{\mathbf{v}_1, \ldots, \mathbf{v}_n\}$. *Then* J *is a face of* P *if and only if either* $J = [n]$ *or* $\mathbf{0} \in \text{relint}(\text{conv}(\{\mathbf{b}_k : k \notin J\}))$.

Proof. Note that $J = [n]$ if and only if J is the whole polytope P which is an improper face of P. So we have to show that $J \subsetneq [n]$ is a face of P if and only if $\mathbf{0} \in \mathrm{relint}(\mathrm{conv}(\{\mathbf{b}_k \; : \; k \notin J\}))$. Let $\dim(P) = d - 1$.

If $J \subsetneq [n]$ is not a face of P, then by Lemma 5.4,

$$\mathrm{aff}(\{\mathbf{v}_k \; : \; k \in J\}) \cap \mathrm{conv}(\{\mathbf{v}_k \; : \; k \notin J\}) \neq \emptyset.$$

Let $\mathbf{z}$ be in this intersection. Then $\mathbf{z} = \sum_{k \in J} p_k \mathbf{v}_k = \sum_{k \notin J} q_k \mathbf{v}_k$ with

$$\sum_{k \in J} p_k = 1, \quad \sum_{k \notin J} q_k = 1, \text{ and } q_k \geq 0 \text{ for all } k \notin J$$

or, equivalently,

$$\sum_{k \in [n]} r_k \mathbf{v}_k = \mathbf{0}, \quad \sum_{k \in [n]} r_k = 0 \text{ and } \sum_{k \notin J} r_k = 1, \; r_k \geq 0 \text{ for all } k \notin J$$

by taking $r_k = q_k$ when $k \notin J$ and $r_k = -p_k$ when $k \in J$.

The first two conditions imply that $\mathbf{r} = (r_1, \ldots, r_n)$ lies in $\ker_{\mathbb{R}}(A)$ where

$$A := \begin{pmatrix} 1 & 1 & \cdots & 1 \\ \mathbf{v}_1 & \mathbf{v}_2 & \cdots & \mathbf{v}_n \end{pmatrix}.$$

Let B be the matrix from which the Gale transform $\mathcal{B}$ was taken. Since the columns of B form a basis for $\ker_{\mathbb{R}}(A)$, there exists $\mathbf{t} \in \mathbb{R}^{n-d}$ such that

$$\mathbf{r} = B\mathbf{t} \text{ or, equivalently, } r_k = \mathbf{b}_k \cdot \mathbf{t} \text{ for all } k = 1, \ldots, n.$$

Since $r_k \geq 0$ for all $k \notin J$, we get that $r_k = \mathbf{b}_k \cdot \mathbf{t} \geq 0$ for all $k \notin J$, which means that all the $\mathbf{b}_k$'s with $k \notin J$ lie in the halfspace defined by $\mathbf{t} \cdot \mathbf{x} \geq 0$ in $\mathbb{R}^{n-d}$. Also since $\sum_{k \notin J} r_k = 1$, it cannot be that $r_k = \mathbf{b}_k \cdot \mathbf{t} = 0$ for all $k \notin J$ or, in other words, not all the $\mathbf{b}_k$'s with $k \notin J$ lie in the hyperplane defined by $\mathbf{t} \cdot \mathbf{x} = 0$. Thus $\mathbf{0}$ is not in the relative interior of $\mathrm{conv}(\{\mathbf{b}_k \; : \; k \notin J\})$. Reversing all the arguments, you get the other direction of the theorem. $\qquad\square$

Example 5.7. Let's use Theorem 5.6 to read off the face lattice of the triangular prism from the Gale diagram in Figure 1. First, note that for each $i = 1, \ldots, 6$, $\mathbf{0} \in \mathrm{relint}(\mathrm{conv}(\mathbf{b}_k \; : \; k \neq i))$. This implies that all the singletons $1, 2, 3, 4, 5, 6$ are faces of P, as indeed they are. Now let's find the edges of P. These will be all pairs

ij such that $\mathbf{0} \in \text{relint}(\text{conv}(\{\mathbf{b}_k : k \neq i,j\}))$. For instance 14 is an edge of P since $\mathbf{0} \in \text{relint}(\text{conv}(\{\mathbf{b}_2, \mathbf{b}_3, \mathbf{b}_5, \mathbf{b}_6\}))$. However, 16 is not an edge of P since $\mathbf{0} \notin \text{relint}(\text{conv}(\{\mathbf{b}_2, \mathbf{b}_3, \mathbf{b}_4, \mathbf{b}_5\}))$. Can you find all the other edges? The face 123 is witnessed by the fact that $\mathbf{0} \in \text{relint}(\text{conv}(\{\mathbf{b}_4, \mathbf{b}_5, \mathbf{b}_6\}))$, but 245 is not a face since $\mathbf{0} \notin \text{relint}(\text{conv}(\{\mathbf{b}_1, \mathbf{b}_3, \mathbf{b}_6\}))$.

Exercise 5.8. Compute the face lattice of the cyclic polytope in $\mathbb{R}^4$ with seven vertices. The Gale transform consists of the columns of the matrix

$$\begin{pmatrix} -1 & 5 & -10 & 10 & -5 & 1 & 0 \\ -5 & 24 & -45 & 40 & -15 & 0 & 1 \end{pmatrix}.$$

(**Hint:** For a simplicial polytope, it suffices to know the facets to write down the whole face lattice.)

Theorem 5.6 can be used to read off the face lattice of any polytope. But it is most useful when the Gale diagram is in a low-dimensional space such as $\mathbb{R}$ or $\mathbb{R}^2$. Three-dimensional Gale diagrams are already quite challenging. However, there is a nice trick to reduce the dimension of the Gale diagram by one. These Gale diagrams are known as *affine Gale diagrams*. See [**Zie95**] for a formal definition. We give the idea below.

We can think of a Gale diagram in $\mathbb{R}^{n-d}$ as n vectors that poke out through a $(n-d-1)$-sphere. If we look at this sphere from outside, we only see one hemisphere, which we will call the northern hemisphere. We can mark all the vectors that poke out through this hemisphere with a dot and label them as before. The rest of the vectors poke out through the southern hemisphere and we will mark their antipodal vectors on the northern hemisphere with an open circle and change labels to the old labels with bars on top. You should always choose the equator so that no vector pokes out through the equator. This can always be done since there are only finitely many vectors in the Gale diagram. Let's first try this on the Gale diagram from Figure 1.

We first put a circle (1-sphere) around the Gale diagram with the dotted line chosen to be the equator. See Figure 4. Let's declare the right hemisphere to be the northern hemisphere. The Gale vectors $1, 6, 2$ intersect this hemisphere. We mark those points with black

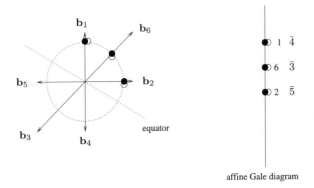

Figure 4. The affine Gale diagram of the triangular prism.

dots. The antipodals of the other vectors also intersect the northern hemisphere at the same points. We mark those intersections with open circles and label them $\bar{4}, \bar{3}, \bar{5}$. On the right we see the affine Gale diagram, which lives in $\mathbb{R}$. Can we read off the face lattice from this affine Gale diagram? To do this, we need to say what condition on a collection of black and white dots is equivalent to the origin being in the relative interior of the Gale vectors with the same indices. For instance to check whether 1346 is a face of P, we remove the dots with labels 1, $\bar{3}$, $\bar{4}$ and 6. This leaves the black dot 2 and the white dot $\bar{5}$ which are at the same position. This means that $\mathbf{b}_2$ and $\mathbf{b}_5$ are opposite to each other and $\mathbf{0}$ is in the relative interior of their convex hull. Thus 1346 is a face.

Exercise 5.9. What conditions on a collection of black and white dots in the affine Gale diagram guarantees that the origin is in the relative interior of the corresponding Gale vectors?

Exercise 5.10. Compute the face lattice of the cyclic polytope in $\mathbb{R}^3$ with seven vertices.

In this case,

$$
A = \begin{pmatrix}
1 & 1 & 1 & 1 & 1 & 1 & 1 \\
1 & 2 & 3 & 4 & 5 & 6 & 7 \\
1 & 4 & 9 & 16 & 25 & 36 & 49 \\
1 & 8 & 27 & 64 & 125 & 216 & 343
\end{pmatrix}.
$$

Using a computer package that does linear algebra, we compute a
basis for $\ker_{\mathbb{R}}(A)$ to get

$$B^t = \begin{pmatrix} 1 & -4 & 6 & -4 & 1 & 0 & 0 \\ 4 & -15 & 20 & -10 & 0 & 1 & 0 \\ 10 & -36 & 45 & -20 & 0 & 0 & 1 \end{pmatrix}.$$

Let's try to draw the affine Gale diagram for this example. We
can start by positioning the last three vectors at the corners of an
equilateral triangle that will be in the center of the hemisphere we
can see. In our case then, we are looking at the sphere along the
vector $(1, 1, 1)$ toward the origin. Can you finish and write down
the face lattice? (**Hint**: Read the rest of this page for a methodical
procedure.)

As the above exercise shows, it is hard to draw affine Gale diagrams precisely, with the description we have of it so far. We need a
more methodical procedure for drawing them, which we now describe.

Let $\mathcal{B} \subset \mathbb{R}^{n-d}$ be the Gale transform. Choose a vector $\mathbf{y} \in \mathbb{R}^{n-d}$
such that $\mathbf{y} \cdot \mathbf{b} \neq 0$ for any $\mathbf{b} \in \mathcal{B}$. We now compute $\mathbf{b}' := \frac{\mathbf{b}}{\mathbf{b} \cdot \mathbf{y}}$ for each
$\mathbf{b} \in \mathcal{B}$. Then the points $\mathbf{b}'$ lie on the hyperplane $H := \{\mathbf{x} \in \mathbb{R}^{n-d} :
\mathbf{y} \cdot \mathbf{x} = 1\}$. If $\mathbf{b}_i \cdot \mathbf{y} > 0$, then label $\mathbf{b}'$ with i and mark it with a black
dot. If $\mathbf{b} \cdot \mathbf{y} < 0$, then label $\mathbf{b}'$ with $\bar{i}$ and mark it with a white dot.
Since H is isomorphic to $\mathbb{R}^{n-d-1}$, we simply have to find an explicit
isomorphism that will help us draw our new points on $H \subset \mathbb{R}^{n-d}$ in
$\mathbb{R}^{n-d-1}$. Projection of the points onto the first $n - d - 1$ coordinates
turns out to be such an isomorphism in the examples you will see in
these chapters.

Exercise 5.11. Compute the face lattice of the four-dimensional
cross-polytope $C^\Delta(4)$ by drawing its affine Gale diagram.

Exercise 5.12. Now replace the vertex $\mathbf{e}_1 \in \mathbb{R}^4$ in $C^\Delta(4)$ with $\alpha \cdot \mathbf{e}_1$.
For different values of $\alpha \in \mathbb{R}$, how will this new convex polytope
change? How is this change reflected in the affine Gale diagram?

Chapter 6

Bizarre Polytopes

In this chapter we will see that Gale diagrams are powerful tools for studying polytopes beyond their ability to encode the faces of a polytope. Let us first investigate some properties of Gale diagrams. The most fundamental question you can ask is if any vector configuration can be the Gale diagram of some polytope. The material in this chapter is taken from [**Zie95**, Chapter 6].

As in Chapter 5, let $V := \{\mathbf{v}_1, \ldots, \mathbf{v}_n\} \subset \mathbb{R}^{d-1}$ and let

$$A = \begin{pmatrix} 1 & \cdots & 1 \\ \mathbf{v}_1 & \cdots & \mathbf{v}_n \end{pmatrix} \in \mathbb{R}^{d \times n}.$$

Assume that $\operatorname{rank}(A) = d$, and choose a matrix $B \in \mathbb{R}^{n \times (n-d)}$ whose columns form a basis of $\ker_{\mathbb{R}}(A)$. Recall that the Gale transform $\mathcal{B} = \{\mathbf{b}_1, \ldots, \mathbf{b}_n\} \subset \mathbb{R}^{n-d}$ consists of the rows of B.

Definition 6.1. (1) A vector configuration $\{\mathbf{w}_1, \ldots, \mathbf{w}_p\} \subset \mathbb{R}^q$ is said to be **acyclic** if there exists a vector $\alpha \in \mathbb{R}^q$ such that $\alpha \cdot \mathbf{w}_i > 0$ for all $i = 1, \ldots, p$. Geometrically this means that all the vectors $\mathbf{w}_i$ lie in the interior of a halfspace defined by a hyperplane in $\mathbb{R}^q$ containing the origin.

(2) A vector configuration $\{\mathbf{w}_1, \ldots, \mathbf{w}_p\} \subset \mathbb{R}^q$ is said to be **totally cyclic** if there exists a vector $\beta > \mathbf{0}$ in $\mathbb{R}^p$ such that $\beta_1 \mathbf{w}_1 + \ldots + \beta_p \mathbf{w}_p = \mathbf{0}$. Geometrically this means that the

---

47

$\mathbf{w}_i$ are arranged all the way around the origin and are not entirely on one side of any hyperplane through the origin.

Lemma 6.2. *The columns of A form an acyclic configuration in $\mathbb{R}^d$ since they all lie in the open halfspace $\{\mathbf{x} \in \mathbb{R}^d : x_1 > 0\}$, while the Gale transform $\mathcal{B}$ is a totally cyclic configuration in $\mathbb{R}^{n-d}$ since $\mathbf{b}_1 + \ldots + \mathbf{b}_n = \mathbf{0}$. (Note that the first row of A, which is a row of ones, dots to zero with the rows of B.)*

Suppose we start with a totally cyclic vector configuration $\mathcal{B} = \{\mathbf{b}_1, \ldots, \mathbf{b}_n\} \subset \mathbb{R}^{n-d}$ and a vector $\beta > \mathbf{0}$ such that $\sum \beta_i \mathbf{b}_i = \mathbf{0}$. By rescaling the elements of $\mathcal{B}$, we may assume that $\beta = (1, 1, \ldots, 1)$. If $B \in \mathbb{R}^{n \times (n-d)}$ is the matrix whose rows are the elements of $\mathcal{B}$, then we can also assume that $\text{rank}(B) = n - d$. This means that $\ker_\mathbb{R}(B^t) = \{\mathbf{x} \in \mathbb{R}^n : B^t \mathbf{x} = \mathbf{0}\}$ is a linear subspace of rank $n - (n - d) = d$. Let $A \in \mathbb{R}^{d \times n}$ be a matrix whose rows form a basis for $\ker_\mathbb{R}(B^t)$. The columns of A form the vector configuration $\mathcal{A} = \{\mathbf{a}_1, \ldots, \mathbf{a}_n\} \subset \mathbb{R}^d$. Then $\mathcal{A}$ is said to be a **Gale dual** of $\mathcal{B}$ and $\mathcal{B}$ a Gale dual of $\mathcal{A}$. By our assumption that $\beta = (1, \ldots, 1)$, we may assume that the first row of A is a row of all ones or, in other words, $\mathbf{a}_i = \begin{pmatrix} 1 \\ \mathbf{v}_i \end{pmatrix}$ for all $i = 1, \ldots, n$. Now the question is, what conditions on $\mathcal{B}$ will ensure that $\{\mathbf{v}_1, \ldots, \mathbf{v}_n\}$ is the vertex set of a $(d-1)$-polytope? To state our answer formally, we introduce the notion of *circuits* and *co-circuits* of vector configurations.

Definition 6.3.   (1) The **sign** of a vector $\mathbf{u} \in \mathbb{R}^n$ is the vector $\text{sign}(\mathbf{u}) \in \{+, 0, -\}^n$ defined as

$$\text{sign}(\mathbf{u})_i := \begin{cases} + & \text{if } u_i > 0 \\ - & \text{if } u_i < 0 \\ 0 & \text{if } u_i = 0. \end{cases}$$

(2) The **support** of a vector $\mathbf{u} \in \mathbb{R}^n$ is the set

$$\text{supp}(\mathbf{u}) := \{i : u_i \neq 0\} \subseteq [n].$$

Note that the supports of a collection of vectors can be partially ordered by set inclusion.

Definition 6.4. Let $\mathcal{W} = \{\mathbf{w}_1, \ldots, \mathbf{w}_p\} \subset \mathbb{R}^q$ be a vector configuration.

(1) A **circuit** of $\mathcal{W}$ is any non-zero vector $\mathbf{u} \in \mathbb{R}^p$ of minimal support such that $\mathbf{w}_1 u_1 + \ldots + \mathbf{w}_p u_p = \mathbf{0}$. The vector sign($\mathbf{u}$) is called a **signed circuit** of $\mathcal{W}$.

(2) A **co-circuit** of $\mathcal{W}$ is any non-zero vector of minimal support of the form $(\mathbf{v} \cdot \mathbf{w}_1, \ldots, \mathbf{v} \cdot \mathbf{w}_n)$ where $\mathbf{v} \in \mathbb{R}^q$. The sign vector of a co-circuit is called a **signed co-circuit**.

Example 6.5. Consider the vector configuration shown in Figure 1 that is the Gale transform of the triangular prism from Chapter 5. If we take $\mathbf{v} = (1, 0)$ in Definition 6.4(2), then we get the co-circuit $(0, 1, -1, 0, -1, 1)$ and the signed co-circuit $(0, +, -, 0, -, +)$. On the other hand, the vector $(1, 0, 0, 1, 0, 0)$ is a circuit of the configuration, and hence $(+, 0, 0, +, 0, 0)$ is a signed circuit of the configuration.

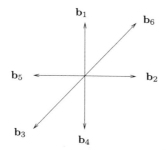

Figure 1. Gale diagram of the vertices of the triangular prism from Chapter 5.

The signed circuits (or, equivalently, signed co-circuits) of a vector configuration completely determine the combinatorics of the configuration. In fact there is a very rich theory of circuits and co-circuits that we will not get into here. It is also a fact that if $\mathcal{A}$ and $\mathcal{B}$ are Gale duals, then the circuits of $\mathcal{A}$ are exactly the co-circuits of $\mathcal{B}$ and vice versa. For instance, in our triangular prism example, $\mathcal{A}$ can be taken to be the columns of the matrix

$$A = \begin{pmatrix} 1 & 1 & 1 & 1 & 1 & 1 \\ 0 & 1 & 0 & 0 & 1 & 0 \\ 0 & 0 & 1 & 0 & 0 & 1 \\ 0 & 0 & 0 & 1 & 1 & 1 \end{pmatrix}.$$

The co-circuit $(0, 1, -1, 0, -1, 1)$ of $\mathcal{B}$ does indeed form a circuit of $\mathcal{A}$: Check first that this vector lies in the kernel of A. To see that it is a circuit, i.e., a dependency on the columns of A of minimal support, you have to check that all subsets of columns $2, 3, 5, 6$ are in fact linearly independent.

Circuits and co-circuits come in symmetric pairs: the negative of a circuit is again a circuit and similarly for co-circuits. It suffices to record one member of each pair.

Theorem 6.6. ([**Zie95**, Theorem 6.19]) *Let* $\mathcal{B} = \{\mathbf{b}_1, \ldots, \mathbf{b}_n\} \subset \mathbb{R}^{n-d}$ *be a totally cyclic vector configuration with* $\sum \mathbf{b}_i = \mathbf{0}$ *and the matrix* B *having rank* $n - d$ *as before. Then* $\mathcal{B}$ *is a Gale transform of a* $(d - 1)$-polytope with n *vertices if and only if every co-circuit of* $\mathcal{B}$ *has at least two positive coordinates.*

Proof. Recall the matrix A constructed from $\mathcal{B}$ as before. We have to show that $\{\mathbf{v}_1, \ldots, \mathbf{v}_n\}$ is the vertex set of the $(d - 1)$-polytope $P = \text{conv}(\{\mathbf{v}_1, \ldots, \mathbf{v}_n\})$ if and only if every co-circuit of $\mathcal{B}$ has at least two positive coordinates. Since $\mathcal{B}$ is totally cyclic, every co-circuit of $\mathcal{B}$ has at least one positive entry and one negative entry. Some co-circuit of $\mathcal{B}$ has exactly one positive entry — say in position j — if and only if $\mathbf{0} \notin \text{relint}(\text{conv}(\mathbf{b}_i : i \neq j))$ which, by Theorem 5.6, is if and only if $\mathbf{v}_j$ is not a vertex of P. This proves the theorem. $\square$

Remark 6.7. If every $\mathbf{v}_i$ is a vertex of $\text{conv}(\{\mathbf{v}_1, \ldots, \mathbf{v}_n\})$, then we say that the $\mathbf{v}_i$ are in **convex position**. Theorem 6.6 provides conditions on a vector configuration $\mathcal{B}$ with the stated assumptions that precisely guarantee when $\mathcal{B}$ is the Gale dual of a configuration $\mathcal{A}$ whose columns are all in convex position.

We can also characterize affine Gale diagrams by reinterpreting Theorem 6.6.

Corollary 6.8. ([**Zie95**, Corollary 6.20]) *A point configuration* $\mathcal{C} = \{\mathbf{c}_1, \ldots, \mathbf{c}_n\} \subset \mathbb{R}^{n-d-1}$, *each of them declared to be either black or white, that affinely spans* $\mathbb{R}^{n-d-1}$, *is the affine Gale diagram of a* $(d - 1)$-polytope with n *vertices if and only if the following condition is satisfied: for every oriented hyperplane* H *in* $\mathbb{R}^{n-d-1}$ *spanned by some points of* $\mathcal{C}$, *the number of black dots on the positive side of* H

plus the number of white dots on the negative side of H is at least two.

Exercise 6.9. Check that Corollary 6.8 is a straight translation of Theorem 6.6 to affine Gale diagrams.

Exercise 6.10. Check that the condition of Corollary 6.8 is true for the affine Gale diagram of the triangular prism from Chapter 5.

We are now ready to get to the fun. We could try to use Gale diagrams to classify $(d-1)$-polytopes with n vertices. Any $(d-1)$-polytope with d vertices is a simplex. The Gale diagram in this case is in zero-dimensional space $\mathbb{R}^0$ and all the $\mathbf{b}_i = 0 \in \mathbb{R}^0$. If P is a $(d-1)$-polytope with $d+1$ vertices, then its Gale diagram is a totally cyclic vector configuration in $\mathbb{R}$ and its affine Gale diagram is a cloud of black and white points in $\mathbb{R}^0$. It is known that there are $\lfloor (d-1)^2/4 \rfloor$ combinatorial types of $(d-1)$-polytopes with $d+1$ vertices. Of these, $\lfloor (d-1)/2 \rfloor$ are simplicial polytopes and the others are multiple pyramids over simplicial polytopes of this type. This is a non-obvious but classical result. Further results are known. See [**Grü03**, Chapter 6] for details. Our goal in the rest of the chapter will be to show that Gale diagrams exhibit the existence of some really bizarre polytopes.

Theorem 6.11. ([**Grü03**, page 94]) *There exists a non-rational eight-dimensional polytope with twelve vertices.*

Proof. Using Corollary 6.8, check that the point configuration shown in Figure 2 is the affine Gale diagram of an 8-polytope P with twelve vertices. It turns out that this point configuration cannot be realized by rational coordinates without violating the prescribed combinatorics. By "prescribed combinatorics" we mean that the same points should be collinear, or on a plane, etc. as in Figure 2. First note that fixing the combinatorics implies that there will always be a pentagon in the middle of the configuration. It is harder to see that this pentagon has to be regular (do you see it?). Furthermore, a regular pentagon cannot be embedded in the plane with rational coordinates as its coordinates will involve $\sqrt{5}$, which is not rational.

Let Q be any polytope that is combinatorially equivalent to P. Then the affine Gale diagram of Q also has the same combinatorics, i.e., same collinearities, circuits, coincidences, etc. Thus Q cannot be realized with rational coordinates either. In particular, neither P nor any polytope combinatorially equivalent to it can be realized with rational coordinates.

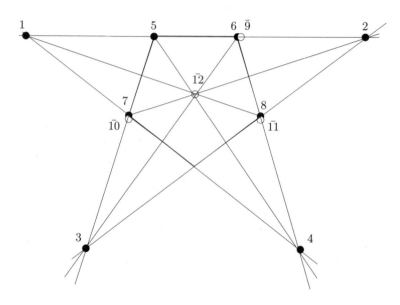

Figure 2. Affine Gale diagram for Theorem 6.11.

□

The above example is due to Perles. No non-rational polytope with less than twelve vertices is known. However, Richter-Gebert has constructed 4-polytopes (with about thirty vertices) which are non-rational. This stands in contrast to the fact that all polytopes of dimension at most three can be realized with rational coordinates. Also, all $(d-1)$-polytopes with at most $d+2$ vertices can be realized with rational coordinates. Can you see how to construct infinitely many polytopes of dimension $d \geq 8$ and at least $d+4$ vertices that do not have rational realizations, beginning with the above example?

We now turn to a different feature of polytopes that can be uncovered via their Gale diagrams. It is known that for all polytopes of dimension $d \leq 3$ or with at most $d + 3$ vertices, one can prescribe the shape of a facet. This means that if a particular facet is known to be an octahedron, say, then we can start with any embedding of an octahedron as this facet and then complete the construction of the polytope according to the combinatorics prescribed. This contrasts the following theorem whose proof is from [**Zie95**, Theorem 6.22].

Theorem 6.12. ([**Stu88**]) *There is a 4-polytope P with seven facets for which the shape of a facet cannot be prescribed.*

Proof. Let P^Δ be the bi-pyramid over a square pyramid. Let the A matrix for this be

$$A = \begin{pmatrix} 1 & 1 & 1 & 1 & 1 & 1 & 1 \\ 1 & 0 & -1 & 0 & 0 & 0 & 0 \\ 0 & 1 & 0 & -1 & 0 & 0 & 0 \\ 0 & 0 & 0 & 0 & 1 & 1 & 0 \\ 0 & 0 & 0 & 0 & 0 & 1 & -1 \end{pmatrix}.$$

To see that the convex hull of the columns of A is a bi-pyramid over a square pyramid, first note that the convex hull of the first four columns of A is a square, and the convex hull of the first five columns of A is a square pyramid. Next note that the average of the last two columns of A is $(1, 0, 0, 1/2, 0)$, which is the midpoint of the line segment perpendicular to the base of the pyramid, dropped from the apex of the pyramid. The columns of

$$\begin{pmatrix} 1 & 0 & 1 & 0 & 2 & -2 & -2 \\ 0 & 1 & 0 & 1 & 2 & -2 & -2 \end{pmatrix}$$

form the Gale transform $\mathcal{B}$ of P^Δ. The Gale diagram is shown in Figure 3.

Now we examine an operation on polytopes that we have not seen so far. The **vertex figure** of a polytope Q at a vertex v is the intersection of Q with a hyperplane H that "chops off" vertex v very near vertex v — i.e., v lies on one side of H and all other vertices of Q lie on the other side of H. The resulting polytope is denoted as Q/v.

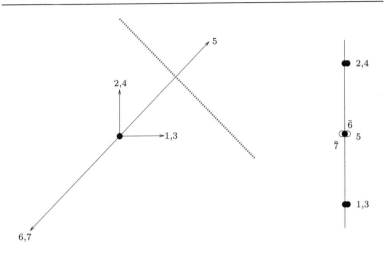

Figure 3. Gale and affine Gale diagram for Theorem 6.12.

Let's consider the vertex figure of our bi-pyramid P^Δ at the vertex 5. The Gale diagram of $P^\Delta/5$ is obtained from the Gale diagram of P^Δ by deleting the point 5 from the diagram. (This is Exercise 6.13.) The resulting Gale diagram is that of a regular octahedron, for instance the one with

$$A' = \begin{pmatrix} 1 & 1 & 1 & 1 & 1 & 1 \\ 1 & -1 & 0 & 0 & 0 & 0 \\ 0 & 0 & 1 & -1 & 0 & 0 \\ 0 & 0 & 0 & 0 & 1 & -1 \end{pmatrix}.$$

Check that for P^Δ, $5\bar{6}$ and $5\bar{7}$ are co-facets. However, this requires that the points $\bar{6}$ and $\bar{7}$ coincide in the affine Gale diagram of $P^\Delta/5$ or, equivalently, that the Gale vectors 6 and 7 are in the linear span of the Gale vector 5 in the opposite direction from 5. Therefore, if we start with a non-regular octahedron such as the following one with

$$A' = \begin{pmatrix} 1 & 1 & 1 & 1 & 1 & 1 \\ 1 & -1 & 0 & 0 & 0 & 0 \\ 1/6 & 0 & 1 & -1 & 0 & 0 \\ 0 & 0 & 0 & 0 & 1 & -1 \end{pmatrix},$$

which has the affine Gale diagram shown in Figure 4, then it is not

Figure 4. Affine Gale diagram of a non-regular octahedron.

the vertex figure of a 4-polytope that is combinatorially isomorphic to P^Δ.

Recall that the face lattices of P and P^Δ are anti-isomorphic, which means that the vertex 5 of P^Δ corresponds to a facet of P. This facet has the same combinatorics as the vertex figure $P^\Delta/5$. Thus by showing that a vertex figure of P^Δ cannot be prescribed, we have shown that a facet of P cannot be prescribed. □

Exercise 6.13. Argue that the Gale diagram of the vertex figure $P^\Delta/5$ is obtained from the Gale diagram of P^Δ by deleting the point 5 from the diagram. This result is true in general.

Chapter 7

Triangulations of Point Configurations

In this chapter we will consider subdivisions and triangulations of graded point configurations. See the book [**DRS**] for a comprehensive account of triangulations in the sense that we will study them.

Let $\mathcal{V} = \{\mathbf{v}_1, \ldots, \mathbf{v}_n\} \subset \mathbb{Z}^{d-1}$ be a point configuration whose convex hull is a $(d-1)$-dimensional polytope $P \subset \mathbb{R}^{d-1}$. We do not insist that the points be distinct or that they all be vertices of P. We embed $\mathcal{V}$ in $\mathbb{R}^d$ by placing all its points on the hyperplane $x_1 = 1$ in $\mathbb{R}^d$ and consider the point configuration

$$\mathcal{A} = \left\{ \begin{pmatrix} 1 \\ \mathbf{v}_1 \end{pmatrix}, \ldots, \begin{pmatrix} 1 \\ \mathbf{v}_n \end{pmatrix} \right\}.$$

The configuration $\mathcal{A}$ is said to be **graded** since it lives on the hyperplane $x_1 = 1$. If $\mathcal{A}$ is graded, then the vector $(1, 1, \ldots, 1)$ lives in the row space of A. Let A be the corresponding $d \times n$ matrix. We have that $\mathrm{rank}(A) = d$. Note that $\mathrm{conv}(\mathcal{A})$ is a $(d-1)$-polytope living in $\mathbb{R}^d$. We will study subdivisions of $\mathcal{A}$.

Once we fix $\mathcal{A}$, we can simply refer to its ith element $\mathbf{a}_i$ by its index $i \in [n]$. We identify $\sigma \subseteq [n]$ with $\mathcal{A}_\sigma := \{\mathbf{a}_i : i \in \sigma\}$. A subset $\mathcal{A}_\sigma$ of a point configuration $\mathcal{A}$ is called a **face** of $\mathcal{A}$ if the elements of $\mathcal{A}_\sigma$ are precisely those that lie on a face of the polytope $\mathrm{conv}(\mathcal{A})$. We refer to this face as σ. The dimension $\dim(\sigma)$ is the dimension of the polytope $\text{conv}(\mathcal{A}_\sigma)$. We say that σ is a k-**simplex** if $\dim(\sigma) = k$ and $|\sigma| = k + 1$. This is all a bit tricky as we will see in the examples in Figure 1.

Definition 7.1. A **subdivision**, $\Delta = \{\sigma_1, \ldots, \sigma_t\}$, of the point configuration $\mathcal{A}$ is a collection of subsets $\sigma_i \subseteq [n]$, $i = 1, \ldots, t$, such that

(1) $\dim(\sigma_i) = d - 1$ for all $i = 1, \ldots, t$,

(2) $\bigcup_{\sigma_i \in \Delta} \text{conv}(\mathcal{A}_{\sigma_i}) = \text{conv}(\mathcal{A})$,

(3) for $i \neq j$, $\text{conv}(\mathcal{A}_{\sigma_i}) \cap \text{conv}(\mathcal{A}_{\sigma_j}) = \text{conv}(\mathcal{A}_\tau)$ where $\tau = \sigma_i \cap \sigma_j$ is a common face of both σ_i and σ_j.

- If, furthermore, all the σ_i are $(d - 1)$-simplices, then Δ is a **triangulation** of $\mathcal{A}$.

- The sets $\{\sigma_i : i = 1, \ldots, t\}$ are called the **facets** $((d - 1)$-faces) of Δ, and the indices that appear in the facets of Δ are called the **vertices** (0-faces) of Δ. Faces of σ_i are called **faces** of Δ.

- A triangulation in which every $i \in [n]$ is a vertex is called a **fine** triangulation of $\mathcal{A}$.

- A subdivision $\Delta = \{\sigma_1, \ldots, \sigma_t\}$ **refines** a subdivision $\Delta' = \{\tau_1, \ldots, \tau_s\}$ if for every $\tau_i \in \Delta'$ there exist $\sigma_{i_1}, \ldots, \sigma_{i_k} \in \Delta$ such that $\{\sigma_{i_1}, \ldots, \sigma_{i_k}\}$ is a subdivision of τ_i.

Example 7.2. Consider the 8-point configuration $\mathcal{A}$ shown in Figure 1. Figure 1(a) is not a subdivision of $\mathcal{A}$ as the union of the facets (which are shaded in this picture) is not all of P. In the rest of the figures, facets will not be shaded. Figure 1(b) is not a subdivision of $\mathcal{A}$ because the facets $\{1, 2, 3\}$ and $\{1, 2, 4\}$ do not intersect in a common face. How about Figure 1(c)? The bottom triangle could mean that either $\{5, 6, 7, 8\}$ is a facet or that $\{6, 7, 8\}$ is a facet. If $\{5, 6, 7, 8\}$ is a facet, then it has the face $\{5, 6, 8\}$ but not $\{5, 8\}$ which means that $\{1, 5, 8\}$ and $\{5, 6, 7, 8\}$ do not intersect in a common face. If $\{6, 7, 8\}$ is a facet, then again $\{1, 5, 8\}$ and $\{6, 7, 8\}$ do not intersect in a common face either. Thus Figure 1(c) is not a subdivision. This example also shows that it is not always clear from the picture what the facets are. *Subdivisions are accurately specified by listing their facets as subsets of* $[n]$.

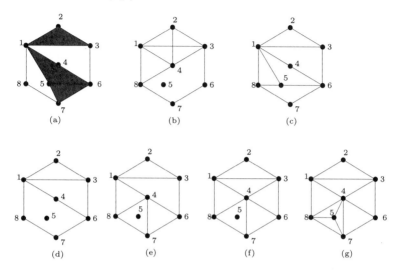

Figure 1. Subdivisions and non-subdivisions.

In Figure 1(d), let us assume that the facets are

$$\{1,2,3\}, \{1,3,4,6\} \text{ and } \{1,4,6,7,8\}.$$

This is indeed a subdivision of $\mathcal{A}$. Some non-faces of this subdivision are $\{5\}$ and $\{1,4\}$. Figure 1(f) is a refinement of Figure 1(d) and is a triangulation. However, Figure 1(e) is not a refinement of Figure 1(d). In fact it is not a subdivision as the facets $\{1,3,4,6\}$ and $\{1,4,8\}$ do not intersect in a common face. How about Figure 1(g)? Hopefully, you can see that it is a triangulation, but is it a refinement of the triangulation in Figure 1(f)? The answer is no since it has simplices that use the vertex 5 which is not present in any face of Figure 1(f).

Let us pause and consider the differences between subdivisions of point configurations as defined above and subdivisions of polytopes that we saw in Chapter 4. For subdivisions of point configurations, all vertices have to come from the configuration. Also, the convex hull of a face may not specify the face as in Figure 1(d) in which the convex hull of $\{1,3,4,6\}$ equals the convex hull of $\{1,3,6\}$.

Among all subdivisions, there are some special ones called *regular subdivisions*, a concept we saw earlier for subdivisions of polytopes. Let's redefine it more generally for point configurations now.

Definition 7.3. Let $\omega = (\omega_1, \ldots, \omega_n) \in \mathbb{R}^n$ be a weight vector and let $\mathcal{A}$ be a graded point configuration in $\mathbb{R}^d$ as earlier. Consider the "lifted" point configuration

$$\mathcal{A}^\omega = \left\{ \begin{pmatrix} \mathbf{a}_1 \\ \omega_1 \end{pmatrix}, \ldots, \begin{pmatrix} \mathbf{a}_n \\ \omega_n \end{pmatrix} \right\} \subset \mathbb{R}^{d+1}$$

and its convex hull $P^\omega \subset \mathbb{R}^{d+1}$. The "lower faces" of P^ω form a polyhedral complex. Projecting this "lower hull" back onto $\mathcal{A}$ induces a subdivision of $\mathcal{A}$, denoted as Δ_ω and known as the **regular subdivision** of $\mathcal{A}$ with respect to ω. Subdivisions of $\mathcal{A}$ that arise via this construction are precisely the regular subdivisions of $\mathcal{A}$. For a generic ω, Δ_ω is a triangulation of $\mathcal{A}$. [In fact we define ω to be generic (with respect to $\mathcal{A}$) whenever Δ_ω is a triangulation of $\mathcal{A}$.]

Recall that F is a lower face of P^ω if for each $\mathbf{x} \in F$ and $\lambda > 0$, $\mathbf{x} - \lambda \mathbf{e}_{d+1} \notin P^\omega$. Figure 2 shows all the regular triangulations of a graded point configuration in $\mathbb{R}^2$ whose convex hull is a line segment.

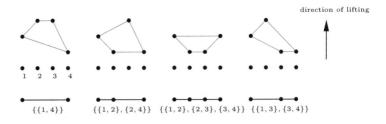

Figure 2. Regular triangulations.

Informally speaking, a subdivision Δ of $\mathcal{A}$ is regular if we can fold it along the creases of the subdivision to be the lower hull of a polytope in one higher-dimensional space. For instance the subdivision in Figure 1(d) is regular as it can be induced by the weight vector $\omega = (0, 5, 1, 0, 100, 0, 0, 0)$. We can test this using PORTA by computing P^ω and computing its lower facets. We will see several alternate tests for regularity in the rest of this chapter. Can you see why the triangulation in Figure 3 is **non-regular** by trying to induce it using a weight vector ω?

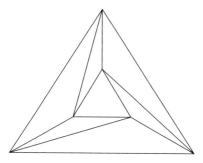

Figure 3. A non-regular triangulation.

The above definition of Δ_ω can be stated mathematically as follows.

Definition 7.4. A subset $\sigma \subseteq [n]$ is a face of the regular subdivision Δ_ω of $\mathcal{A}$ if and only if there exists a vector $\mathbf{y} \in \mathbb{R}^d$ such that

$$\mathbf{a}_j \cdot \mathbf{y} = \omega_j \quad \text{for all } j \in \sigma,$$
$$\mathbf{a}_j \cdot \mathbf{y} < \omega_j \quad \text{for all } j \notin \sigma.$$

What is the above test doing geometrically? It says that σ is a face if and only if the vector $(\mathbf{y}, -1) \cdot (\mathbf{a}_j, \omega_j)^t = 0$ for all $j \in \sigma$ and $(\mathbf{y}, -1) \cdot (\mathbf{a}_j, \omega_j)^t < 0$ for all $j \notin \sigma$. This means that $(\mathbf{y}, -1)$ is normal to the affine hull of the "lifted" σ and all other lifted points lie above this affine hull.

While Definition 7.4 is useful for testing whether $\Delta = \Delta_\omega$, given ω, perhaps we might first ask for a more general test of regularity. Given a subdivision Δ of $\mathcal{A}$, is it regular? In other words, does $\Delta = \Delta_\omega$ for some ω? To proceed, we need to be able to work with *cones* which are special cases of polyhedra.

Definition 7.5. A **cone** $K \subseteq \mathbb{R}^d$ is any subset of $\mathbb{R}^d$ such that

(1) for $\mathbf{x}, \mathbf{y} \in K$, $\mathbf{x} + \mathbf{y} \in K$, and

(2) for $\mathbf{x} \in K$ and $\lambda \geq 0$, $\lambda \mathbf{x} \in K$.

The most common example of a cone is the ice-cream cone in $\mathbb{R}^3$, an example of which is the set

$$K = \{(x, y, z) \in \mathbb{R}^3 \ : \ x^2 + y^2 \leq z^2, \ z \geq 0\}.$$

We will be interested in *polyhedral cones* which are cones with finitely many flat sides.

Definition 7.6. A **polyhedral cone** $K \subseteq \mathbb{R}^d$ is a polyhedron of the form

$$K = \{\mathbf{x} \in \mathbb{R}^d \ : \ M\mathbf{x} \geq \mathbf{0}\}$$

where M is a real matrix or, equivalently (by the main theorem of cones), any set of the form

$$K = \{N\mathbf{y} \ : \ \mathbf{y} \geq \mathbf{0}\}$$

where N is a real matrix.

The second expression says that K is the set of all non-negative combinations of columns of N. The columns $\mathcal{N}$ of N are called the generators of the cone K. This can be expressed by writing K as $K = \text{cone}(\mathcal{N})$. A column $\mathbf{n}$ of N generates an **extreme ray** of K if $\text{cone}(\mathcal{N}) \supsetneq \text{cone}(\mathcal{N}\backslash\mathbf{n})$. The inequalities in $M\mathbf{x} \geq \mathbf{0}$ are called the constraints of K. By the main theorem of cones, which is analogous to the main theorem for polytopes, every finitely constrained cone is a finitely generated cone and vice versa. A **cone complex** is a polyhedral complex in which all the polyhedra are cones. Cone complexes are more typically called **polyhedral fans**. If the support of the fan is the entire space it lives in, we call it a **complete fan**.

Definition 7.7. Let $P \subset \mathbb{R}^d$ be a polyhedron and let F be a face of P. Then the **outer normal cone** of P at F is the cone

$$\mathcal{N}_P(F) = \{\mathbf{c} \in \mathbb{R}^d \ : \ F = \text{face}_{\mathbf{c}}(P)\}.$$

The collection of outer normal cones of P is called the **outer normal fan** of P. It is denoted as $\mathcal{N}(P)$. Similarly the **inner normal cone** of P at F is the cone

$$\mathcal{N}_P(F) = \{\mathbf{c} \in \mathbb{R}^d \ : \ F = \text{face}_{-\mathbf{c}}(P)\}$$

and the **inner normal fan** of P is the fan formed by the collection of inner normal cones of P. We use the same notation for both as we will not need both simultaneously anywhere in this book.

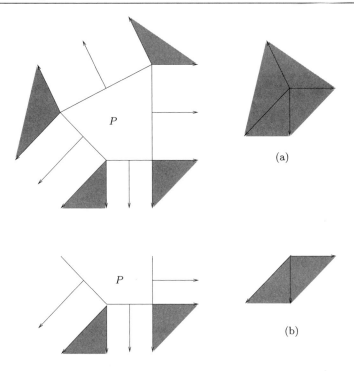

Figure 4. A complete and incomplete fan.

Figure 4 gives two examples of outer normal fans of polyhedra. The top fan is complete while the bottom fan is not complete.

Notice that the outer normal cone of P at a face F is the negative of the inner normal cone of P at F. Also, if a face F is contained in a face G of P, then $\mathcal{N}_P(F)$ contains $\mathcal{N}_P(G)$. The map that sends a face F of P to its normal cone $\mathcal{N}_P(F)$ is an anti-isomorphism. For instance, if P is a full-dimensional polyhedron in $\mathbb{R}^d$, then the normal cone at a k-face of P is a $(d-k)$-face of $\mathcal{N}(P)$.

Exercise 7.8. Prove that both the inner and outer normal fans of a polyhedron are cone complexes. (You will need to show that the intersection of two cones in a fan is a common face of each and that every face of every cone in a fan is again a cone in the fan.)

Definition 7.9. A polyhedral fan is **polytopal** if it is the outer/inner normal fan of a polyhedron.

A very surprising fact is that not all polyhedral fans are outer normal fans of polyhedra! To see such an example, take the fan whose cones are the cones over the simplices in the non-regular triangulation in Figure 3. Argue that this fan is not polytopal.

Recall that in Chapter 5 we defined the Gale transform of a configuration $\mathcal{V}$ by first grading it to form the matrix A and then taking the kernel of A. If V is already graded (we are given A) we do not need to grade it again to compute the Gale transform. In this case, we say that $\mathcal{B}$ is a Gale transform of $\mathcal{A}$ ($= \mathcal{V}$).

We now return to the question of how we can check whether a given subdivision Δ of a point configuration $\mathcal{A}$ is regular. For a set $\sigma \subseteq [n]$, let $\bar{\sigma} := [n] \backslash \sigma$.

Theorem 7.10. ([**Lee91**]) *Let* $\Delta = \{\sigma_1, \ldots, \sigma_t\}$ *be a subdivision of* $\mathcal{A}$ *and let* $\mathcal{B}$ *be a Gale transform of* $\mathcal{A}$. *Then* Δ *is regular if and only if*

$$\bigcap_{i=1}^{t} \mathrm{relint}(\mathrm{cone}(\mathcal{B}_{\bar{\sigma}_i})) \neq \emptyset.$$

The proof of this theorem will be the main job remaining. Before we do that, let's understand the result geometrically. First, note that the test involves figuring out whether the relative interiors of the cones $\{\mathrm{cone}(\mathcal{B}_{\bar{\sigma}_i})\}$ have a common intersection. Since each $\sigma_i \in \Delta$ has at least d elements, $\mathcal{B}_{\bar{\sigma}_i}$ has at most $n - d$ elements and $\mathrm{cone}(\mathcal{B}_{\bar{\sigma}_i}) \subset \mathbb{R}^{n-d}$. Let's first apply this theorem to check the regularity of the triangulations in Figure 2.

Example 7.11. Suppose the configuration $\mathcal{A}$ shown in Figure 2 is the grading of $\{0, 1, 2, 3\}$. Then

$$A = \begin{pmatrix} 1 & 1 & 1 & 1 \\ 0 & 1 & 2 & 3 \end{pmatrix} \text{ and } B^t = \begin{pmatrix} 1 & -2 & 1 & 0 \\ 2 & -3 & 0 & 1 \end{pmatrix}.$$

The Gale diagram is shown in Figure 5(a).

(1) $\Delta_1 = \{\{1, 4\}\}$ is regular since $\mathrm{relint}(\mathrm{cone}(\{\mathbf{b}_2, \mathbf{b}_3\}))$ is non-empty.

(2) $\Delta_2 = \{\{1, 2\}, \{2, 4\}\}$ is regular since $\mathrm{relint}(\mathrm{cone}(\{\mathbf{b}_3, \mathbf{b}_4\})) \cap \mathrm{relint}(\mathrm{cone}(\{\mathbf{b}_1, \mathbf{b}_3\}))$ is $\mathrm{relint}(\mathrm{cone}(\{\mathbf{b}_1, \mathbf{b}_3\}))$, which is not empty.

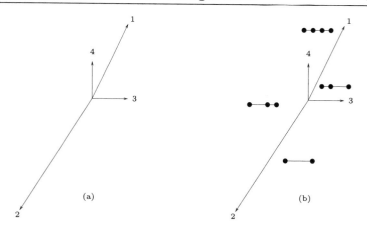

Figure 5. Gale diagram of the configuration in Figure 2.

(3) $\Delta_3 = \{\{1, 2\}, \{2, 3\}, \{3, 4\}\}$ is regular since
$\mathrm{relint}(\mathrm{cone}(\{\mathbf{b}_3, \mathbf{b}_4\})) \cap \mathrm{relint}(\mathrm{cone}(\{\mathbf{b}_1, \mathbf{b}_4\})) \cap$
$\mathrm{relint}(\mathrm{cone}(\{\mathbf{b}_1, \mathbf{b}_2\}))$ is $\mathrm{relint}(\mathrm{cone}(\{\mathbf{b}_1, \mathbf{b}_4\}))$, which is not
empty.

(4) $\Delta_4 = \{\{1, 3\}, \{3, 4\}\}$ is regular since $\mathrm{relint}(\mathrm{cone}(\{\mathbf{b}_2, \mathbf{b}_4\})) \cap$
$\mathrm{relint}(\mathrm{cone}(\{\mathbf{b}_1, \mathbf{b}_2\}))$ is $\mathrm{relint}(\mathrm{cone}(\{\mathbf{b}_2, \mathbf{b}_4\}))$, which is not
empty.

In Figure 5(b) we have placed the four regular triangulations in
the corresponding $\bigcap_{i=1}^{t} \mathrm{relint}(\mathrm{cone}(\mathcal{B}_{\bar{\sigma}_i}))$.

Example 7.12. Consider the graded point configuration $\mathcal{A}$ consisting
of the columns of the matrix

$$\begin{pmatrix} 4 & 0 & 0 & 2 & 1 & 1 \\ 0 & 4 & 0 & 1 & 2 & 1 \\ 0 & 0 & 4 & 1 & 1 & 2 \end{pmatrix}$$

and the subdivisions of $\mathcal{A}$ shown in Figure 6.

The Gale transform of $\mathcal{A}$ consists of the columns of the matrix

$$\begin{pmatrix} 1 & 0 & 0 & -3 & 1 & 1 \\ 0 & 1 & 0 & 1 & -3 & 1 \\ 0 & 0 & 1 & 1 & 1 & -3 \end{pmatrix}$$

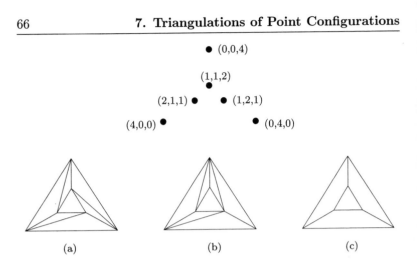

Figure 6.

and the affine Gale diagram is shown in Figure 7. (We used the construction described at the end of Chapter 5 with $\mathbf{y} = (1,1,1)$.)

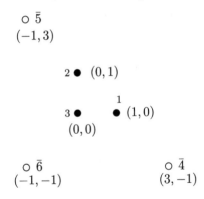

Figure 7. Affine Gale diagram of the configuration in Example 7.12.

The subdivision in Figure 6(c) is $\Delta_3 = \{\sigma_1 = \{4,5,6\}, \sigma_2 = \{1,2,4,5\}, \sigma_3 = \{2,3,5,6\}, \sigma_4 = \{1,3,4,6\}\}$. Figure 8 shows the cones $\{\mathrm{cone}(\mathcal{B}_{\bar{\sigma}_i})\}$. Note that their relative interiors intersect at the point $\mathbf{z}$, which shows that Δ_3 is regular.

Exercise 7.13. Check that Figure 6(b) is a regular triangulation while Figure 6(a) is non-regular.

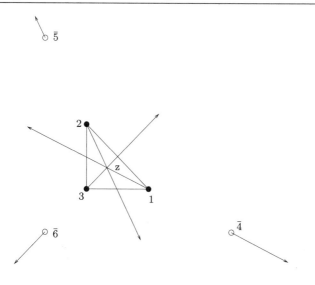

Figure 8. Intersection of $\{\text{cone}(\mathcal{B}_{\bar{\sigma}_i}) : \sigma_i \in \Delta_3\}$.

If $\mathcal{B}$ is a Gale transform of a point configuration $\mathcal{A}$, then the scaled vector configuration $\tilde{\mathcal{B}} := \{\mu_1 \mathbf{b}_1, \ldots, \mu_n \mathbf{b}_n\}$ where $\mu_i > 0$ is called a **scaled** Gale transform of $\mathcal{A}$. Scaled Gale transforms carry the same combinatorial information as Gale transforms — for any $\sigma \subsetneq [n]$, $\mathbf{0} \in \text{relint}(\text{conv}(\mathcal{B}_{\bar{\sigma}}))$ if and only if $\mathbf{0} \in \text{relint}(\text{conv}(\tilde{\mathcal{B}}_{\bar{\sigma}}))$. Let's prove one direction: $\mathbf{0} \in \text{relint}(\text{conv}(\mathcal{B}_{\bar{\sigma}}))$ if and only if $\mathbf{0} = \sum_{i \in \bar{\sigma}} \lambda_i \mathbf{b}_i$ where $0 < \lambda_i < 1$, $\sum_{i \in \bar{\sigma}} \lambda_i = 1$. Suppose $\sum_{i \in \bar{\sigma}} \frac{\lambda_i}{\mu_i} = t$. Then clearly $t > 0$ since $\lambda_i, \mu_i > 0$. Then $\mathbf{0} = \sum_{i \in \bar{\sigma}} \frac{\lambda_i}{t \mu_i}(\mu_i \mathbf{b}_i)$ where $0 < \frac{\lambda_i}{t \mu_i} < 1$ and $\sum_{i \in \bar{\sigma}} \frac{\lambda_i}{t \mu_i} = 1$, which shows that $\mathbf{0} \in \text{relint}(\text{conv}(\tilde{\mathcal{B}}_{\bar{\sigma}}))$. Can you prove the other direction? The upshot is that we can read off the faces of $P = \text{conv}(\mathcal{A})$ from $\tilde{\mathcal{B}}$. In many books, all scaled Gale transforms are called Gale diagrams.

Now we begin to prove Theorem 7.10 via a series of lemmas. We have to show that $\Delta = \{\sigma_1, \ldots, \sigma_t\}$ is a regular subdivision of $\mathcal{A}$ if and only if

$$\bigcap_{i=1}^{t} \text{relint}(\text{cone}(\mathcal{B}_{\bar{\sigma}_i})) \neq \emptyset.$$

The strategy will be to show that elements in $\bigcap_{i=1}^{t} \mathrm{relint}(\mathrm{cone}(\mathcal{B}_{\bar{\sigma}_i}))$ are precisely the *seeds* for lifting vectors $\omega \in \mathbb{R}^n$ such that $\Delta = \Delta_\omega$. This proof is an elaboration of the one in [**Lee91**].

Lemma 7.14. *Let $\omega = (\omega_1, \ldots, \omega_n) \in \mathbb{R}^n$ such that $0 < \omega_i < 1$ and consider the polytopes*

$$P^\omega = \mathrm{conv}\left(\left\{\begin{pmatrix} \mathbf{a}_1 \\ \omega_1 \end{pmatrix}, \ldots, \begin{pmatrix} \mathbf{a}_n \\ \omega_n \end{pmatrix}\right\}\right) \text{ and}$$

$$P^{\mathbf{1}-\omega} = \mathrm{conv}\left(\left\{\begin{pmatrix} \mathbf{a}_1 \\ (1-\omega_1) \end{pmatrix}, \ldots, \begin{pmatrix} \mathbf{a}_n \\ (1-\omega_n) \end{pmatrix}\right\}\right).$$

Then the facets in the lower hull of P^ω are precisely the facets in the upper hull of $P^{\mathbf{1}-\omega}$.

Exercise 7.15. Prove the above lemma. Figure 9 shows a small example that might make the lemma believable. The configuration $\mathcal{A}$ is on the $x_{d+1} = 0$ plane. The polytope P^ω has unshaded vertices while $P^{\mathbf{1}-\omega}$ has shaded vertices.

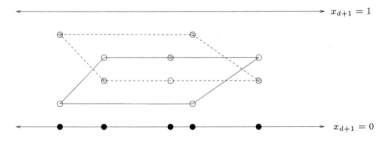

Figure 9. Figure for Lemma 7.14.

Lemma 7.16. *Let $\mathcal{B} \subset \mathbb{R}^{n-d}$ be a Gale transform of the graded point configuration $\mathcal{A} = \{\mathbf{a}_1, \ldots, \mathbf{a}_n\} \subset \mathbb{R}^d$. Let $\mathbf{z}$ be a vector in the interior of the convex hull of $\mathcal{B}$ and let $0 < \omega_1, \ldots, \omega_n < 1$, $\sum_{i=1}^{n} \omega_i = 1$ be such that $\mathbf{z} = \sum_{i=1}^{n} \omega_i \mathbf{b}_i$. Then $\mathcal{B}' := \mathcal{B} \cup \{-\mathbf{z}\}$ is a scaled Gale transform of the vector configuration*

$$\mathcal{A}' = \left\{\frac{1}{\omega_1}\begin{pmatrix} \mathbf{a}_1 \\ (1-\omega_1) \end{pmatrix}, \ldots, \frac{1}{\omega_n}\begin{pmatrix} \mathbf{a}_n \\ (1-\omega_n) \end{pmatrix}, \begin{pmatrix} \mathbf{0} \\ -1 \end{pmatrix}\right\}.$$

Proof. To prove this, we have to show that $\mathcal{B}'$ is obtained by positively scaling some Gale transform of $\mathcal{A}'$. Using the definition of $\mathbf{z}$ and Gale transforms, check that the matrix product

$$
\left(\frac{1}{\omega_1} \begin{pmatrix} \mathbf{a}_1 \\ (1 - \omega_1) \end{pmatrix} \quad \cdots \quad \frac{1}{\omega_n} \begin{pmatrix} \mathbf{a}_n \\ (1 - \omega_n) \end{pmatrix} \begin{pmatrix} \mathbf{0} \\ -1 \end{pmatrix} \right) \begin{pmatrix} \omega_1 \mathbf{b}_1 \\ \vdots \\ \omega_n \mathbf{b}_n \\ -\mathbf{z} \end{pmatrix}
$$

gives the zero matrix. This proves that the columns of the second matrix lie in the kernel of A'. Since $\mathcal{B}$ consists of $n - d$ linearly independent vectors, the rank of the second matrix is $n - d$, which is the dimension of the kernel of A'. This proves that the rows of the second matrix form a Gale transform of $\mathcal{A}'$. Clearly $\mathcal{B}'$ is a positive scaling of this Gale transform. $\qquad\square$

Proof of Theorem 7.10. To finish the proof, we work out two equivalent descriptions of the facets of $\mathrm{conv}(\mathcal{A}')$ that do not contain the point $(\mathbf{0}, -1)$. Consider the projective transformation

$$
f : \mathbb{R}^{d+1} \to \mathbb{R}^{d+1} \text{ such that } \begin{pmatrix} \mathbf{x} \\ x_{d+1} \end{pmatrix} \mapsto \frac{1}{1 + x_{d+1}} \begin{pmatrix} \mathbf{x} \\ x_{d+1} \end{pmatrix}.
$$

Under this map,

$$
\frac{1}{\omega_i} \begin{pmatrix} \mathbf{a}_i \\ (1 - \omega_i) \end{pmatrix} \mapsto \begin{pmatrix} \mathbf{a}_i \\ (1 - \omega_i) \end{pmatrix}
$$

and $(\mathbf{0}, -1)^t$ gets sent to the plane at infinity. This implies that $\mathrm{conv}(\mathcal{A}')$ gets sent to $P^{1-\omega} + [\mathbf{0}, -\infty]$ and the facets of $\mathrm{conv}(\mathcal{A}')$ that do not contain $(\mathbf{0}, -1)$ are precisely the bounded facets of $P^{1-\omega} + [\mathbf{0}, -\infty]$ which in turn are the facets in the upper hull of $P^{1-\omega}$.

On the other hand, we can use $\mathcal{B}'$ to work out the facets of $\mathrm{conv}(\mathcal{A}')$ that do not contain the point $(\mathbf{0}, -1)$. Recall that σ is such a facet if it is maximal with the property that

$$
\mathbf{0} \in \mathrm{relint}(\mathrm{conv}(\mathcal{B}_{\bar{\sigma}} \cup \{-\mathbf{z}\})).
$$

Now notice that we can rephrase this last sentence as "σ is such a facet if it is maximal with the property that $\mathbf{z} \in \mathrm{relint}(\mathrm{cone}(\mathcal{B}_{\bar{\sigma}}))$".

Thus we have proved that the facets in the upper hull of $P^{1-\omega}$ (lower hull of P^ω) are precisely the maximal sets $\sigma \subseteq [n]$ with

$$\bigcap_{i=1}^{t} \text{relint}(\text{cone}(\mathcal{B}_{\bar{\sigma}_i})) \neq \emptyset.$$

$\square$

Remark 7.17. The proof of Theorem 7.10 also shows us how to produce weight vectors ω that induce a regular triangulation Δ. By the theorem,

$$\bigcap_{\sigma \in \Delta} \text{relint}(\text{cone}(\mathcal{B}_{\bar{\sigma}})) \neq \emptyset.$$

Pick a non-zero $\mathbf{z}$ in this intersection small enough such that $\mathbf{z} \in \text{relint}(\text{conv}(\mathcal{B}))$. Then there exists $\omega = (\omega_1, \ldots, \omega_n) \in \mathbb{R}^n$ such that $\mathbf{z} = \sum_{i=1}^{n} \omega_i \mathbf{b}_i$, $0 < \omega_i < 1$, $\sum_{i=1}^{n} \omega_i = 1$, and $\Delta = \Delta_\omega$.

Note that once we have ω, any positive multiple of it will also induce the same subdivision. Thus we could have picked any $\mathbf{z} \in \bigcap_{\sigma \in \Delta} \text{relint}(\text{cone}(\mathcal{B}_{\bar{\sigma}}))$ and solved for $\mathbf{z} = \omega B$ to get ω such that $\Delta = \Delta_\omega$.

Exercise 7.18. Find weight vectors to induce the four regular triangulations in Figure 5. How about the subdivision Δ_3 in Figure 6 (c)?

Exercise 7.19. Calculate all the regular subdivisions of the 4-point configuration from Figure 2 and draw them in the picture of the Gale diagram in Figure 5(a). Note that in Figure 5(b) we have found all the regular triangulations. They correspond to the four full-dimensional cones that you see in the natural fan induced by the Gale diagram.

Definition 7.20. If a subdivision Δ' refines a subdivision Δ of a point configuration $\mathcal{A}$, then write $\Delta' \succ \Delta$. The operation $\succ$ partially orders the set $\mathcal{S}$ of all subdivisions of $\mathcal{A}$. This poset $(\mathcal{S}, \succ)$ is called the **refinement poset** of the subdivisions of $\mathcal{A}$.

Exercise 7.21. Describe the refinement poset of the 4-point configuration in Figure 2.

Exercise 7.22. Let $\mathcal{A}$ be the point configuration with six points in $\mathbb{R}^3$ given in Example 7.12. Carefully describe its refinement poset.

Indicate which subdivisions Δ are regular by giving a weight vector ω that induces each regular subdivision.

Chapter 8

The Secondary Polytope

Theorem 7.10 from the last chapter tests for the regularity of a subdivision Δ of a graded point configuration $\mathcal{A} \subset \mathbb{Z}^d$ with the stated assumptions. This theorem suggests an algorithm for constructing all regular triangulations of $\mathcal{A}$. Take the Gale diagram $\mathcal{B} \subset \mathbb{R}^{n-d}$ of $\mathcal{A}$ and a vector $\mathbf{z} \in \mathbb{R}^{n-d}$. Then the subdivision

$$\Delta = \{\sigma \subseteq [n] \ : \ \mathbf{z} \in \text{relint}(\text{cone}(\mathcal{B}_{\bar{\sigma}}))\}$$

is a regular subdivision of $\mathcal{A}$. Conversely, if Δ is regular, then

$$\bigcap_{i=1}^{t} \text{relint}(\text{cone}(\mathcal{B}_{\bar{\sigma}_i})) \neq \emptyset$$

and we can choose a $\mathbf{z}$ in this intersection.

In the last chapter, we also saw that if $\mathbf{z} = \sum_{i=1}^{n} \omega_i \mathbf{b}_i$, then, in fact, $\Delta = \Delta_\omega$. Is this ω unique? Recall that the elements of $\mathcal{B}$ are the rows of an $n \times (n-d)$ matrix B whose columns form a basis for $\ker_{\mathbb{R}}(A)$. Then $\mathbf{z} = \sum_{i=1}^{n} \omega_i \mathbf{b}_i = \omega B$. Let $\mathbf{y} \in \text{row space}(A)$. Then $\mathbf{y}B = \mathbf{0}$. This implies that $\mathbf{z} = \omega B + \mathbf{y}B = (\omega + \mathbf{y})B$. Thus we can add any vector in row space(A) to a solution ω to the linear system $\mathbf{z} = \omega B$ to get ω' such that $\Delta_\omega = \Delta_{\omega'}$. In particular, since $\mathcal{A}$ is assumed to be graded, $(1, 1, \ldots, 1) \in \text{row space}(A)$, and there is always a positive weight vector ω such that $\mathbf{z} = \omega B$. Any ω such that $\mathbf{z} = \omega B$ is said to be "lifted" from $\mathbf{z}$.

Definition 8.1.   (1) The **secondary cell** of a regular subdivision Δ of $\mathcal{A}$ is the open cone

$$\{\omega \; : \; \text{there exists } \mathbf{z} = \omega B, \, \mathbf{z} \in \bigcap_{i=1}^{t} \text{relint}(\text{cone}(\mathcal{B}_{\bar{\sigma}_i}))\} \subseteq \mathbb{R}^n.$$

   (2) The closure of the secondary cell in $\mathbb{R}^n$ is the **secondary cone** of Δ, denoted as $\mathcal{C}_\Delta$.

   (3) The open cone $\bigcap_{i=1}^{t} \text{relint}(\text{cone}(\mathcal{B}_{\bar{\sigma}_i})) \subset \mathbb{R}^{n-d}$ is the **pointed secondary cell** of Δ, and

   (4) the closure of $\bigcap_{i=1}^{t} \text{relint}(\text{cone}(\mathcal{B}_{\bar{\sigma}_i}))$ is called the **pointed secondary cone** of Δ, which we denote as $\mathcal{C}'_\Delta$.

Note that the pointed secondary cell and cone lie in $\mathbb{R}^{n-d}$ and hence do not consist of height/weight vectors that induce triangulations. To get weight vectors ω, we have to solve the system $\mathbf{z} = \omega B$ for $\mathbf{z}$ in these pointed cones. In other words, we can think of B as defining a linear transformation from $\mathbb{R}^n \to \mathbb{R}^{n-d}$ where $\omega \mapsto \omega B$. The full-dimensional cones are the preimages under this map of the corresponding pointed cones. The pointed cones allow us to mod out the *lineality space*, row space(A), from the set of height functions and hence are more convenient to study. The **lineality space** of a cone is the largest subspace in the cone.

The discussion before Definition 8.1 proves that every vector in the secondary cell of Δ is a weight vector that induces Δ as a regular subdivision. Conversely, if $\omega \in \mathbb{R}^n$ such that $\Delta = \Delta_\omega$, then we can write ω as $\omega = \omega' + \omega''$ where $\omega' \in \ker_{\mathbb{R}}(A)$ and $\omega'' \in$ row space(A). The proof of Theorem 7.10 shows that $\mathbf{z} = \omega B = \omega' B$ lies in $\bigcap_{i=1}^{t} \text{relint}(\text{cone}(\mathcal{B}_{\bar{\sigma}_i}))$. This proves the following theorem.

Theorem 8.2. *The secondary cell of a regular subdivision Δ is precisely the set of all weight vectors $\omega \in \mathbb{R}^n$ such that $\Delta = \Delta_\omega$.*

Example 8.3. Consider the vector configuration $\mathcal{A}$ consisting of the vertices of the pentagon shown in Figure 1. Here

$$A = \begin{pmatrix} 1 & 1 & 1 & 1 & 1 \\ 0 & 1 & 2 & 1 & 0 \\ 0 & 0 & 1 & 2 & 1 \end{pmatrix} \text{ and } B = \begin{pmatrix} -2 & -2 \\ 3 & 2 \\ -2 & -1 \\ 1 & 0 \\ 0 & 1 \end{pmatrix}.$$

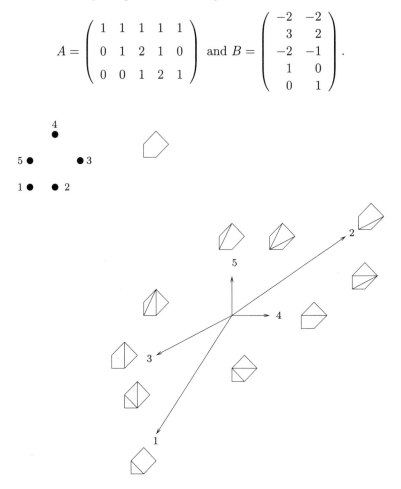

Figure 1. The regular subdivisions of a pentagon.

Figure 1 shows all the regular subdivisions of $\mathcal{A}$ placed against their pointed secondary cones. The origin is the pointed secondary cone of the coarsest subdivision $\Delta_0 = \{\{1, 2, 3, 4, 5\}\}$ shown at the top of the figure. Its secondary cone is the subspace row space(A) in $\mathbb{R}^5$. The five regular triangulations of $\mathcal{A}$ have full-dimensional secondary cones. For instance, the triangulation

$$\Delta = \{\{1,2,3\}, \{3,4,5\}, \{1,3,5\}\}$$

has the pointed secondary cone, cone($\{\mathbf{b}_2, \mathbf{b}_4\}$), which gives a five-dimensional secondary cone. The remaining five regular subdivisions have one-dimensional pointed secondary cones which give four-dimensional secondary cones in $\mathbb{R}^5$.

Example 8.3 might make you suspect that there is much more structure to the regular subdivisions of a point configuration than what we have so far. For instance, we have the following.

(1) The secondary cones of the regular subdivisions seem to form a polyhedral fan with the full-dimensional cones in the fan indexed by regular triangulations.

(2) The face lattice of this fan seems to correspond to the poset of regular subdivisions of $\mathcal{A}$ ordered by refinement. See Figure 2.

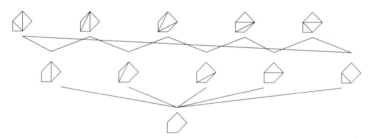

Figure 2. The refinement poset of the subdivisions of the pentagon.

(3) There seems to be a relationship between a regular subdivision and its neighbors in this refinement poset. In particular, the two triangulations neighboring a subdivision seem to be the two possible regular triangulations that refine this subdivision.

(4) Lastly, in this example at least, it seems that the fan of secondary cones is polytopal. There is a pentagon whose normal fan is this fan.

All these observations are in fact theorems! We focus on (1) and (4) and see examples and construction methods.

Consider the Gale transform $\mathcal{B}$ as a vector configuration in $\mathbb{R}^{n-d}$. We call $\tau = \{\tau_1, \ldots, \tau_{n-d}\} \subset [n]$ a **basis** if $\mathrm{cone}(\mathcal{B}_\tau)$ is a basis of $\mathbb{R}^{n-d}$.

Definition 8.4. ([**BFS90**]) The **pointed secondary fan** of $\mathcal{A}$, denoted as $\mathcal{F}'(\mathcal{A})$, is the cone complex obtained as the multi-intersection of all the cones $\{\mathrm{cone}(\mathcal{B}_\tau)\}$ as τ ranges over all bases of $\mathcal{B}$.

The **multi-intersection** of a collection of cones is the new cone complex consisting of all intersections of the original cones involved. Do not confuse it with the common intersection of all the cones involved, which may be just the origin.

Each cell (open cone) in the complex is of the form

$$\bigcap_{\sigma \in \Delta} \mathrm{relint}(\mathrm{cone}(\mathcal{B}_{\bar{\sigma}}))$$

where Δ consists of all the sets $\sigma \subseteq [n]$ that are maximal with the property that the cell lies in $\mathrm{relint}(\mathrm{cone}(\mathcal{B}_{\bar{\sigma}}))$. Then Δ is the regular subdivision indexed by this cell. In particular, the full-dimensional cells in $\mathcal{F}'(\mathcal{A})$ index the regular triangulations of $\mathcal{A}$ since if a cell is full-dimensional, then the maximal σ's such that the cell lies in $\mathrm{relint}(\mathrm{cone}(\mathcal{B}_{\bar{\sigma}}))$ all have d elements.

Example 8.5. Let us verify all this in Example 8.3. Looking at Figure 1, we see that the multi-intersection of $\mathrm{cone}(\mathcal{B}_\tau)$ as τ ranges over all bases of $\mathcal{B}$ is precisely the fan obtained by simply drawing the two-dimensional Gale transform in the plane. Therefore, the pointed secondary fan $\mathcal{F}'(A)$ is the two-dimensional fan shown in Figure 1. Check that the maximal σ's such that the interior of $\mathrm{cone}(\mathbf{b}_5)$ lies in the relative interior of $\mathrm{cone}(\mathcal{B}_{\bar{\sigma}})$ are $\{1, 2, 3, 4\}$ and $\{1, 4, 5\}$, which are the facets of the subdivision labeling this pointed secondary cell.

Example 8.6. The secondary fan becomes more complicated when $n - d > 2$. Let us try to construct the pointed secondary fan of the configuration $\mathcal{A}$ consisting of the columns of

$$\begin{pmatrix} 4 & 0 & 0 & 2 & 1 & 1 \\ 0 & 4 & 0 & 1 & 2 & 1 \\ 0 & 0 & 4 & 1 & 1 & 2 \end{pmatrix}$$

whose Gale transform $\mathcal{B}$ consists of the columns of

$$\begin{pmatrix} 1 & 0 & 0 & -3 & 1 & 1 \\ 0 & 1 & 0 & 1 & -3 & 1 \\ 0 & 0 & 1 & 1 & 1 & -3 \end{pmatrix}.$$

We saw the Gale diagram in the previous chapter. To construct the pointed secondary fan, we have to find all bases of $\mathcal{B}$ and then intersect the cones they span. How do we do this? Let B_τ be the square $(n-d) \times (n-d)$ submatrix of B whose columns are the elements of $\mathcal{B}_\tau$. Then τ is a basis if and only if B_τ is non-singular. In Figure 3 we have drawn the fan in the northern hemisphere which is the hemisphere in which we had previously drawn the affine Gale diagram. You see fifteen full-dimensional cones in this picture. In the southern hemisphere there is one more which is the cone spanned by $\mathcal{B}_{\{4,5,6\}}$. Thus there are sixteen full-dimensional cells in this pointed secondary fan which implies that this configuration has sixteen regular triangulations.

Let's construct the regular triangulation that goes with the cell marked with an **x**. This cell lies in the relative interior of $\mathrm{cone}(\mathcal{B}_{\bar\sigma})$ for the following $\bar\sigma$'s:

$$\{1,3,4\}, \{2,3,6\}, \{2,3,5\}, \{1,2,3\}, \{3,4,6\}, \{2,4,5\}, \{1,4,5\},$$

which implies that it corresponds to the regular triangulation

$$\Delta = \{\{2,5,6\}, \{1,4,5\}, \{1,4,6\}, \{4,5,6\}, \{1,2,5\}, \{1,3,6\}, \{2,3,6\}\}$$

shown in Figure 4.

Exercise 8.7. Find all the regular triangulations of the above configuration and draw them in their pointed secondary cones.

Remark 8.8. The software package TOPCOM [**Ram**] can be used to find all regular triangulations of a point configuration.

Theorem 8.9. *The face poset of the secondary fan of $\mathcal{A}$ and the refinement poset of $\mathcal{A}$ have isomorphic Hasse diagrams.*

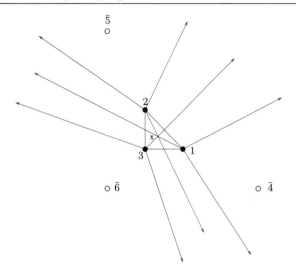

Figure 3. The pointed secondary fan in the northern hemisphere.

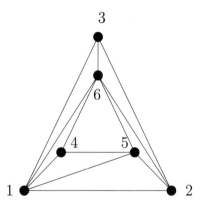

Figure 4. The regular triangulation for the cell marked x.

We will not elaborate on the above theorem except to say that we hope it is somewhat believable based on the examples.

Exercise 8.10. Isolate the part of the pointed secondary fan in Figure 3 contained in the cone spanned by $\mathcal{B}_{\{1,2,3\}}$. Draw the part of the refinement poset of $\mathcal{A}$ indexed by the cells (of all dimensions) that appear in this piece of the pointed secondary fan.

We conclude this chapter by showing that the secondary fan of $\mathcal{A}$ is polytopal. In fact, it is enough to show that the pointed secondary fan is polytopal as we can then embed this polytope in $\ker_{\mathbb{R}}(A) \subset \mathbb{R}^n$ and get its normal fan in $\mathbb{R}^n$ to be the pointed secondary fan plus the row space of A. Recall that we have assumed that $\mathcal{A}$ is a graded point configuration in $\mathbb{Z}^d$.

Definition 8.11. Let σ be a simplex in a triangulation of $\mathcal{A}$. The **normalized volume** of the simplex σ, denoted as $\mathrm{vol}(\sigma)$, is the absolute value of the determinant of A_σ divided by the greatest common divisor (g.c.d.) of the maximal minors of A.

Example 8.12. In the $\mathcal{A}$ of Example 8.6, the g.c.d. of the maximal minors of A is four. Thus the normalized volume of the simplex $\{4, 5, 6\}$ is one while the normalized volume of the simplex $\{1, 2, 5\}$ is four.

Definition 8.13. The **GKZ vector** of a triangulation Δ of $\mathcal{A}$ is the vector

$$\phi_\Delta := \sum_{i=1}^{n} \left(\sum \{\mathrm{vol}(\tau) \, : \, \tau \in \Delta \text{ and } i \in \tau\} \right) \cdot \mathbf{e}_i \in \mathbb{R}^n.$$

GKZ stands for Gel'fand, Kapranov and Zelevinsky who discovered and initiated much of this work [**GKZ94**].

Definition 8.14. A **secondary polytope** of $\mathcal{A}$ is any polytope whose inner normal fan equals the secondary fan $\mathcal{F}(\mathcal{A})$.

Theorem 8.15. ([**GKZ94**]) *The polytope*

$$\Sigma(\mathcal{A}) := \mathrm{conv}(\{\phi_\Delta \, : \, \Delta \text{ a triangulation of } \mathcal{A}\})$$

is a secondary polytope of $\mathcal{A}$. The vertices of $\Sigma(\mathcal{A})$ are the GKZ vectors of the regular triangulations of $\mathcal{A}$.

Proof. ([**BFS90**, §2]) In order to show that $\Sigma(\mathcal{A})$ is a secondary polytope of $\mathcal{A}$, we have to show that $\mathcal{N}(\Sigma(\mathcal{A}))$, the inner normal fan of $\Sigma(\mathcal{A})$, equals $\mathcal{F}(\mathcal{A})$, the secondary fan of $\mathcal{A}$. We will prove that for each regular triangulation Δ of $\mathcal{A}$, the secondary cone $\mathcal{C}_\Delta$ is contained in $\mathcal{N}_{\Sigma(\mathcal{A})}(\phi_\Delta)$. Recall that $\mathcal{N}_{\Sigma(\mathcal{A})}(\phi_\Delta)$ is the inner normal cone of $\Sigma(\mathcal{A})$ at the vertex ϕ_Δ. This will prove the theorem since both $\mathcal{N}(\Sigma(\mathcal{A}))$ and $\mathcal{F}(\mathcal{A})$ are complete polyhedral fans in $\mathbb{R}^n$.

Given a triangulation Δ of $\mathcal{A}$ and a vector $\omega \in \mathbb{R}^n$, we obtain a unique function $g_{\omega,\Delta}$ with domain $P = \mathrm{conv}(\mathcal{A})$ as follows: set $g_{\omega,\Delta}(\mathbf{a}_i) = \omega_i$ for the vertices $\mathbf{a}_i$ of Δ and require that $g_{\omega,\Delta}$ be an affine function on each simplex of Δ. This is an example of a *piecewise linear* function on Δ. When $\Delta = \Delta_\omega$, the graph of $g_{\omega,\Delta}$ is the lower hull of $P^\omega = \mathrm{conv}(\mathcal{A}^\omega)$ where $\mathcal{A}^\omega$ is the lifting of $\mathcal{A}$ by ω.

Let Δ be a regular triangulation of $\mathcal{A}$, and let $\omega \in \mathcal{C}_\Delta$. For any point $\mathbf{a}_j \in \mathcal{A}$, the point $(\mathbf{a}_j, \omega_j)$ lies on or above the graph of $g_{\omega,\Delta}$. So if we take a different triangulation Δ' and consider $g_{\omega,\Delta'}$, then its graph is contained on or above the graph of $g_{\omega,\Delta}$. In other words, $g_{\omega,\Delta}(\mathbf{x}) \leq g_{\omega,\Delta'}(\mathbf{x})$ for all $\mathbf{x} \in P$. This implies that

$$\int_{\mathbf{x}\in P} g_{\omega,\Delta}(\mathbf{x})dx \leq \int_{\mathbf{x}\in P} g_{\omega,\Delta'}(\mathbf{x})dx$$

for all triangulations Δ' of $\mathcal{A}$. Now observe that

$$
\begin{aligned}
\int_{\mathbf{x}\in P} g_{\omega,\Delta}(\mathbf{x})dx &= \sum_{\tau\in\Delta,\,\text{facet}} \int_{\mathbf{x}\in\tau} g_{\omega,\Delta}(\mathbf{x})dx \\
&= \sum_{\tau\in\Delta,\,\text{facet}} \mathrm{vol}(\tau)(\text{barycentric value of}\,g_{\omega,\Delta}\,\text{on}\,\tau) \\
&= \sum_{\tau\in\Delta,\,\text{facet}} \mathrm{vol}(\tau) \cdot \frac{1}{d}\sum_{i\in\tau} g_{\omega,\Delta}(\mathbf{a}_i) \\
&= \frac{1}{d}\sum_{i=1}^{n} \omega_i \sum_{\tau\in\Delta\,:\,i\in\tau} \mathrm{vol}(\tau) \\
&= \frac{1}{d}\,(\omega\cdot\phi_\Delta).
\end{aligned}
$$

Since the same formula holds for Δ', we get that $(\omega\cdot\phi_\Delta) \leq (\omega\cdot\phi_{\Delta'})$ for all triangulations Δ' of $\mathcal{A}$. But this implies that ϕ_Δ lies in $\mathrm{face}_{-\omega}(\Sigma(\mathcal{A}))$ or, equivalently, ω is contained in the inner normal cone $\mathcal{N}_{\Sigma(\mathcal{A})}(\phi_\Delta)$. $\qquad\square$

Remark 8.16. The above proof also shows that the collection of full-dimensional secondary cones form a polyhedral fan — the inner normal fan of $\Sigma(\mathcal{A})$.

Example 8.17. Let us construct the secondary polytope of the configuration in Example 8.3.

regular triangulation	GKZ vector
$\Delta_1 = \{\{1,2,3\}, \{1,3,4\}, \{1,4,5\}\}$	$(5,1,4,4,1)$
$\Delta_2 = \{\{1,2,4\}, \{2,3,4\}, \{1,4,5\}\}$	$(3,4,2,5,1)$
$\Delta_3 = \{1,2,5\}, \{2,3,4\}, \{2,4,5\}\}$	$(1,5,2,4,3)$
$\Delta_4 = \{\{1,2,5\}, \{2,3,5\}, \{3,4,5\}\}$	$(1,3,4,2,5)$
$\Delta_5 = \{\{1,2,3\}, \{1,3,5\}, \{3,4,5\}\}$	$(3,1,5,2,4)$

This can be read off from the determinants of the simplices of this configuration that are computed below:

simplex τ	vol(τ)
$\{1,2,3\}$	1
$\{1,2,4\}$	2
$\{1,2,5\}$	1
$\{1,3,4\}$	3
$\{1,3,5\}$	2
$\{1,4,5\}$	1
$\{2,3,4\}$	2
$\{2,3,5\}$	2
$\{2,4,5\}$	2
$\{3,4,5\}$	2

Computing the convex hull of the five GKZ vectors in PORTA, we get the following description of the secondary polytope which is two-dimensional.

```
INEQUALITIES_SECTION
(   1)        +   x2+ x3-   x4-x5  ==   0
(   2) +13x1+11x2-4x3-15x4        ==   0
(   3)                +  x3+ 2x4+x5 == 13

(   1) -3x4-2x5 <=  -14
```

```
(  2) - x4      <=  -2
(  3)      - x5 <=  -1
(  4) + x4+ x5  <=   7
(  5) +2x4+ x5  <=  11
```

END

```
strong validity table :
\ P      |         |
 \ O     |         |
I \ I    |         |
 N \ N   | 1       | #
  E \ T  |         |
---------------------
1        | *...* :   2
2        | ...** :   2
3        | **... :   2
4        | ..**. :   2
5        | .**.. :   2
         . . . . . . . . . .
#        | 22222
```

This shows that the secondary polytope is a pentagon as we expect and that the fan shown in Figure 1 is its inner normal fan. Verify this.

Exercise 8.18.   (1) Compute the GKZ coordinates of the six regular triangulations in Exercise 8.10 and the GKZ vector of the non-regular triangulation in this family that we have been seeing. Where does the vector of the non-regular triangulation lie in the secondary polytope?

(2) Furthermore, deduce that the vertices of $\Sigma(\mathcal{A})$ corresponding to these regular triangulations lie on a common facet of the secondary polytope. (It seems that we must find the normalized volumes of many triangulations, but really we only have to find it for two of them!)

Chapter 9

The Permutahedron

In this chapter we investigate the secondary polytope of a prism over a simplex. This material is taken entirely from [**DRS**, §5.2].

Recall that $\Delta_n = \mathrm{conv}(\{\mathbf{e}_1, \ldots, \mathbf{e}_{n+1}\})$ is the unit n-simplex and $I = \Delta_1$ is a line segment. We wish to study the prism over an n-simplex, any example of which is combinatorially equivalent to the product $\Delta_n \times I$. Thus we will use $\Delta_n \times I$ to denote a prism over an n-simplex. The prism $\Delta_n \times I$ is an $(n+1)$-dimensional polytope with $2(n + 1)$ vertices. It has two simplicial facets which are both copies of Δ_n which we call the top and bottom facets of $\Delta_n \times I$. In addition it has $n + 1$ vertical facets of the form $\Delta_{n-1} \times I$.

For convenience we will denote the vertices of the bottom simplicial facet by $p_1, \ldots, p_{n+1}$ and the vertices of the top simplicial facet by $q_1, \ldots, q_{n+1}$ such that p_i is directly under q_i. Note that any four vertices of the form p_i, q_i, p_j, q_j form a quadrilateral 2-face of the prism. This is key to proving the following fact about maximal simplices in a triangulation of $\Delta_n \times I$.

Lemma 9.1. *A set of $n + 2$ vertices of $\Delta_n \times I$ form the vertices of an $(n + 1)$-simplex if and only if two of them form a pair p_i, q_i and the remaining n of them are taken one from each of the remaining n pairs $\{p_j, q_j\}$, $i \neq j$.*

Exercise 9.2. Prove Lemma 9.1.

Lemma 9.1 says that if σ is a maximal simplex in a triangulation of $\Delta_n \times I$, then σ has n vertical facets contained in the vertical facets of $\Delta_n \times I$ that are not incident to any other maximal simplex of the triangulation. There are two other facets — one opposite the vertex p_i and the other opposite the vertex q_i. We call these the top and bottom facets of σ, respectively. This implies that every triangulation of $\Delta_n \times I$ has a linearly ordered sequence of simplices, starting with a simplex incident to the top facet of the prism and ending with a simplex incident to the bottom facet of the prism such that the bottom facet of one simplex is incident to the top facet of the next simplex. For an example, see Figure 1.

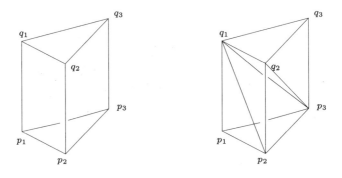

Figure 1. A triangulation of $\Delta_n \times I$.

Theorem 9.3. *There is a bijection between the triangulations of $\Delta_n \times I$ and the permutations of $[n+1]$ which are the elements of the symmetric group S_{n+1}.*

(1) Let $\pi = i_1 \cdots i_{n+1}$ be a permutation of $[n+1]$. Then the following $n+1$ simplices form a triangulation of $\Delta_n \times I$:

$$T_\pi := \left\{ \{p_{i_1}, \ldots, p_{i_k}, q_{i_k}, \ldots, q_{i_{n+1}}\} : k = 1, \ldots, n+1 \right\}.$$

(2) All triangulations of $\Delta_n \times I$ have this form. In particular, they are all equivalent to each other.

(3) Two regular triangulations of $\Delta_n \times I$ are adjacent vertices of the secondary polytope of $\Delta_n \times I$ if and only if the corresponding permutations differ by the exchange of a pair of consecutive elements.

In particular, $\Delta_n \times I$ has exactly $(n+1)!$ triangulations and each triangulation has exactly $n+1$ simplices.

We will not prove (3) as we have not developed the notion of adjacency of triangulations. Two regular triangulations are adjacent if they index adjacent vertices of the secondary polytope. However, this is a special case of a more general notion of adjacency between pairs of triangulations that extends to even the non-regular triangulations and creates a graph in which the triangulations are the vertices and the edges are defined by this notion of adjacency that we are referring to. The edge graph of the secondary polytope is an $(n-d)$-connected subgraph of this graph of all triangulations.

Proof. ([**DRS**]) Let's prove (2). The proof of (1) is similar. Let T be a triangulation of $\Delta_n \times I$ and σ_1 its unique maximal simplex incident to the top facet of the prism. Let $p_{i_1} \in \sigma_1$ be the vertex opposite to the top facet of σ_1. The only facet of σ_1 that is interior to the prism is the bottom facet — opposite to the vertex q_{i_1}. This facet is the top facet of the next maximal simplex σ_2 in T which is obtained by deleting q_{i_1} from σ_1 and inserting a second vertex p_{i_2}. Again the bottom facet of σ_2 (the one opposite q_{i_2}) is the top facet of the next maximal simplex σ_3 containing a third bottom vertex p_{i_3} and so on. The vertex set of the kth simplex we get in this process will be $\{p_{i_1}, \ldots, p_{i_k}, q_{i_k}, \ldots, q_{i_{n+1}}\}$. $\qquad\square$

Example 9.4. (1) $n = 1$: In this case $\Delta_1 \times I$ is a square (identified with C_2). It has precisely two triangulations shown in Figure 2 which are in bijection with the permutations of $\{1, 2\}$. The triangulations are
   (a) $T_{12} := \{\{p_1, q_1, q_2\}, \{p_1, p_2, q_2\}\}$,
   (b) $T_{21} := \{\{p_2, q_2, q_1\}, \{p_2, p_1, q_1\}\}$.

(2) $n = 2$: In this case $\Delta_2 \times I$ is a triangular prism. It has $3! = 6$ triangulations which are shown according to their adjacencies in Figure 2. The secondary polytope is two-dimensional. All six triangulations are regular and thus the secondary polytope is a hexagon. You can check Theorem 9.3(3) on this example. The permutation indexing a triangulation is written on top of the triangulation. Notice

that these permutations are adjacent — i.e., they are related by the exchange of two consecutive letters in each.

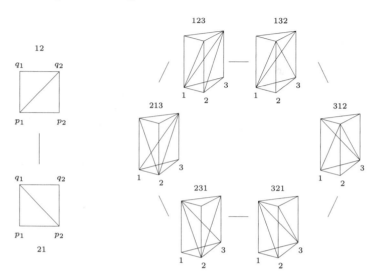

Figure 2. Triangulations of $\Delta_1 \times I$ and $\Delta_2 \times I$.

Exercise 9.5. In the triangulation T_π show that the diagonal in the square with vertices p_i, q_i, p_j, q_j will go from p_i to q_j if and only if i comes before j in the permutation π. Check this on the triangulations in Figure 2. Can you prove this in general? This gives an easy way to draw T_π given π.

Looking carefully at the two secondary polytopes we implicitly have in Figure 2, we see that its vertices are labeled by all the elements of a symmetric group and that two vertices are adjacent when the permutations are adjacent in the sense we have described. Since the symmetric group is a classical object and adjacent permutations have great significance in the study of symmetric groups, the following polytope should come as no surprise.

Definition 9.6. ([**Zie95**, Chapter 0]) Let $\pi = i_1 \cdots i_n \in S_n$ and $\mathbf{v}_\pi := (i_1, \ldots, i_n) \in \mathbb{R}^n$. The **permutahedron** Π_n is the convex hull of the vectors $\{\mathbf{v}_\pi : \pi \in S_n\}$.

The permutahedron Π_n is an $(n-1)$-dimensional polytope in $\mathbb{R}^n$. (Note that all the vectors $\mathbf{v}_\pi$ lie on the hyperplane $\sum x_i = \frac{n(n+1)}{2}$.) Each $\mathbf{v}_\pi$ is a vertex of Π_n and two vertices $\mathbf{v}_\pi$ and $\mathbf{v}_{\pi'}$ are adjacent if and only if the permutations π and π' are adjacent (i.e., differ by the exchange of two consecutive elements). The faces of a permutahedron are products of lower-dimensional permutahedra. Verify this in Figure 3 which shows a Schlegel diagram of Π_4.

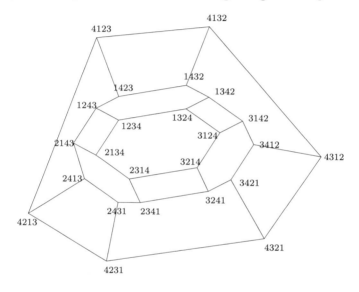

Figure 3. The permutahedron Π_4.

We will now prove that the secondary polytope of $\Delta_n \times I$ is affinely isomorphic to the permutahedron Π_{n+1}. The secondary polytope will have dimension $2(n+1)-(n+1)-1 = n$ and will have $(n+1)!$ vertices, which agrees with the dimension and number of vertices of Π_{n+1}.

Definition 9.7. A triangulation of $\mathcal{A}$ is said to be **unimodular** if all its simplices are unimodular (i.e., have unit normalized volume).

Exercise 9.8. Check that all triangulations of the two prisms in Figure 2 are unimodular.

Lemma 9.9. *All full-dimensional simplices of $\Delta_n \times I$ have the same volume. In particular, all triangulations of $\Delta_n \times I$ are unimodular.*

Proof. We prove this by induction on n. You can check from Figure 2 that the statement is true for $n = 1$. Let F be a vertical facet of $\Delta_n \times I$. Then F is affinely isomorphic to $\Delta_{n-1} \times I$ and by induction, all maximal simplices contained in F have the same volume. The pair of vertices p_i, q_i opposite to F are at the same distance from the affine span of F. Hence any two $(n+1)$-simplices having a facet in F must have their unique other vertex be either p_i or q_i and hence have the same volume. However, every pair of $(n+1)$-simplices of $\Delta_n \times I$ has facets in a common vertical facet of $\Delta_n \times I$, which proves the lemma.   □

Theorem 9.10. *The secondary polytope of $\Delta_n \times I$ is affinely isomorphic to Π_{n+1}.*

Proof. Let T_π be a triangulation of $\Delta_n \times I$ where $\pi = i_1 \cdots i_{n+1} \in S_{n+1}$. Let ϕ_{T_π} be the GKZ vector of T_π and let $\phi_{T_\pi}(p_{i_j})$ be its coordinate corresponding to p_{i_j}. Recall that

$$T_\pi = \left\{ \{p_{i_1}, \ldots, p_{i_k}, q_{i_k}, \ldots, q_{i_{n+1}}\} \ : \ k = 1, \ldots, n+1 \right\}.$$

This implies that if q_{i_j} is in k simplices, then p_{i_j} is in $n + 1 - k + 1 = n + 2 - k$ simplices. Thus, by Lemma 9.9,

$$\phi_{T_\pi}(p_{i_j}) = n + 2 - \phi_{T_\pi}(q_{i_j}) \text{ for all } j = 1, \ldots, n+1.$$

Thus if we know the p_{i_j}-coordinates of the GKZ vector of T_π, we can reconstruct the q_{i_j}-coordinates. This also means that we can project all the GKZ vectors to the $(n+1)$-vectors corresponding to their q-coordinates and get a polytope that is affinely isomorphic to the secondary polytope of $\Delta_n \times I$. We already saw that the vertex q_{i_k} lies in exactly k simplices of T_π, which means that the projected GKZ vector of T_π is precisely $\mathbf{v}_{\pi^{-1}}$ where π^{-1} is the permutation inverse to π in the symmetric group S_{n+1}. Putting all this together, we have that the secondary polytope of $\Delta_n \times I$ is affinely isomorphic to the permutahedron Π_{n+1}.   □

Corollary 9.11. *All triangulations of $\Delta_n \times I$ are regular.*

To end this chapter, we briefly investigate the secondary polytope of an n-gon which is commonly known as an *associahedron*. This material is also taken from [**DRS**], but most of it will be developed through exercises.

Exercise 9.12. Prove that all triangulations of the n-gon are regular. **Hint:** Note that all triangulations of the 4-gon and 5-gon have at least two faces with vertices $\{i, i+1, i+2\}$. Show that this is true for any n-gon.

Exercise 9.13. Show that the number of triangulations of the n-gon is $\frac{1}{n-1}\binom{2n-4}{n-2}$.

Hints: The sequence $(\frac{1}{n+1}\binom{2n}{n})$ is known as the *Catalan sequence* and its terms, the Catalan numbers. For Exercise 9.13 let $\mathcal{T}_n$ denote the set of all triangulations of the n-gon and $t_n := |\mathcal{T}_n|$. Let c_1 be the surjective map $c_1 : \mathcal{T}_{n+1} \longrightarrow \mathcal{T}_n$ that contracts the edge $\{1, n+1\}$ in each triangulation of the $(n+1)$-gon. Proceed as follows:

(1) Given some fixed triangulation $\Delta \in \mathcal{T}_n$, how many triangulations in $\mathcal{T}_{n+1}$ will be mapped to Δ under c_1? Experiment with $n = 4$.

(2) Let $\deg(1, \Delta)$ be the number of edges in Δ that contain the vertex 1. Argue that
$$t_{n+1} = \sum_{\Delta \in \mathcal{T}_n} \deg(1, \Delta).$$

(3) There was nothing special about contracting along the edge $\{1, n+1\}$. Argue that
$$n t_{n+1} = \sum_{i=1}^{n} \sum_{\Delta \in \mathcal{T}_n} \deg(i, \Delta).$$

(4) Argue that for any $\Delta \in \mathcal{T}_n$,
$$\sum_{i=1}^{n} \deg(i, \Delta) = 2(2n - 3).$$

(5) Now give a (multiplicative) recurrence for t_{n+1}. Repeat.

The Catalan numbers count many other well-studied and important combinatorial objects, and we will explore two of these and see how they relate to triangulations of the n-gon. For this we take the 5-gon as our working example.

Definition 9.14. A **binary tree** is a tree where every node has at most two descendants, one labeled L (left) and the other R (right).

Let Δ be a triangulation of the $(n+2)$-gon with facets $\{\sigma_i\}$. Place a vertex u_i in the interior of every σ_i and form the *dual graph* **dual**(Δ) with vertices u_i and edges $\{u_i, u_j\}$ whenever $\sigma_i \cap \sigma_j$ intersect at an edge of Δ. Then dual(Δ) is a binary tree with n nodes and $n - 1$ vertices. Do the following exercises for the 5-gon and then convince yourself that these results are true in general.

Exercise 9.15. Show that there is a bijection between the set $\mathcal{T}_{n+2}$ and the set of binary trees on n vertices.

A *parenthesization* of $n + 1$ objects is any bracketing of the $n + 1$ objects that specifies an order in which to carry out multiplication. For example $(a(bc))$ is a parenthesization of the three objects a, b, c that says to first compute bc and then multiply by a.

Given a binary tree with n nodes, consider the following rule for parenthesizing $n + 1$ objects: *For the root place one pair of brackets around the $n + 1$ objects and recursively perform the following: if the right/left child of a given parent has $k - 1 \geq 0$ descendants, then enclose the $k + 1$ rightmost/leftmost factors within the ones enclosed in the parent parentheses.*

Exercise 9.16. Show that there is a bijection between the set of binary trees on n vertices and the set of parenthesizations of $n + 1$ objects.

Exercise 9.17. From Figure 1 of Chapter 8, confirm that two vertices of the secondary polytope of the 5-gon form an edge of the secondary polytope if their corresponding parentheses differ only by one use of the associative law $a(bc) = (ab)c$. (This is true in general: the edges of the secondary polytope of an n-gon are defined by those vertices that differ by only one use of the associative law. This explains the name *associahedron* for the secondary polytopes of n-gons.)

Chapter 10

Abstract Algebra: Polynomial Rings

In this chapter we will study polynomial rings and ideals which will eventually get us to more polytopes. See [**CLO92**] for more details on this material. Recall that $\mathbb{N}$ denotes the set of non-negative integers.

Definition 10.1. (1) A **monomial** in the n variables $x_1, \ldots, x_n$ is a product of the form $x_1^{a_1} x_2^{a_2} \cdots x_n^{a_n}$ where $a_1, a_2, \ldots, a_n \in \mathbb{N}$. In short form, we write the monomial $x_1^{a_1} x_2^{a_2} \cdots x_n^{a_n}$ as $\mathbf{x}^{\mathbf{a}}$ where $\mathbf{x} = (x_1, \ldots, x_n)$ and $\mathbf{a} = (a_1, \ldots, a_n) \in \mathbb{N}^n$.

(2) A **polynomial** in the n variables $x_1, \ldots, x_n$ over a ring R is a finite linear combination of monomials $\mathbf{x}^{\mathbf{a}}$ with coefficients in R. It has the general form $f(x_1, \ldots, x_n) = f(\mathbf{x}) = \sum c_{\mathbf{a}} \mathbf{x}^{\mathbf{a}}$ where $c_{\mathbf{a}} \in R$. If the set of variables is clear, then we typically denote $f(\mathbf{x})$ as just f.

(3) The simplest polynomials are of the form $c_{\mathbf{a}} \mathbf{x}^{\mathbf{a}}$. They are called **terms**.

(4) The **support** of a polynomial $f = \sum c_{\mathbf{a}} \mathbf{x}^{\mathbf{a}}$ is the finite set $\operatorname{supp}(f) := \{\mathbf{a} \in \mathbb{N}^n : c_{\mathbf{a}} \neq 0\}$.

Let $R[x_1, \ldots, x_n] = R[\mathbf{x}]$ denote the set of all polynomials in $x_1, \ldots, x_n$ over R. We can define a natural addition and multiplication in $R[\mathbf{x}]$ as follows. Given two polynomials $f = \sum c_{\mathbf{a}}^f \mathbf{x}^{\mathbf{a}}$ and

$g = \sum c_{\mathbf{b}}^g \mathbf{x}^{\mathbf{b}}$ in $R[\mathbf{x}]$, let

$$f + g := \sum_{\mathbf{m}\in\mathrm{supp}(f)\cup\mathrm{supp}(g)} (c_{\mathbf{m}}^f + c_{\mathbf{m}}^g)\mathbf{x}^{\mathbf{m}}.$$

In the above sum, if $\mathbf{m} \in \mathrm{supp}(f)\backslash\mathrm{supp}(g)$, then set $c_{\mathbf{m}}^g = 0$ and similarly, if $\mathbf{m} \in \mathrm{supp}(g)\backslash\mathrm{supp}(f)$, then set $c_{\mathbf{m}}^f = 0$. The product of two monomials $\mathbf{x}^{\mathbf{a}}$ and $\mathbf{x}^{\mathbf{b}}$ is the new monomial $\mathbf{x}^{\mathbf{a}+\mathbf{b}}$. Extending this to polynomials, we define the product of the polynomials f and g as

$$f \cdot g := \sum_{\{(\mathbf{a},\mathbf{b})\,:\,\mathbf{a}\in\mathrm{supp}(f),\,\mathbf{b}\in\mathrm{supp}(g)\}} c_{\mathbf{a}}^f c_{\mathbf{b}}^g \mathbf{x}^{\mathbf{a}+\mathbf{b}}.$$

Exercise 10.2. Let $f = x_1^3 x_2^6 - \frac{1}{2}x_3^5 + 16$ and $g = \sqrt{2}x_2^5 + 13$ be polynomials in $\mathbb{R}[x_1, x_2, x_3]$. Find $f + g$ and g^2.

Definition 10.3. The set $R[\mathbf{x}]$ of all polynomials in $x_1, \ldots, x_n$ over R, along with the operations of addition and multiplication defined above, is called the **polynomial ring** in $x_1, \ldots, x_n$ over R. The ring R is called the **coefficient ring** of $R[\mathbf{x}]$.

Exercise 10.4. Check that $R[\mathbf{x}]$ is a ring under the operations of addition and multiplication on polynomials defined above.

Every polynomial in $R[\mathbf{x}]$ has a natural geometric object associated to it called its **zero-set** or **variety**. This is the collection of all points $\{(r_1, \ldots, r_n) \in R^n : f(r_1, \ldots, r_n) = 0\}$ denoted as $V_R(f)$. An element of $V_R(f)$ is called a **zero** or **root** of f. For example, if $x^2 - 1 \in \mathbb{Z}[x]$, then $V_{\mathbb{Z}}(x^2 - 1) = \{1, -1\}$. In fact, $V_{\mathbb{Z}}(x^2 - 1) = V_{\mathbb{R}}(x^2 - 1) = V_{\mathbb{C}}(x^2 - 1)$. However, for the polynomial $x^2 + 1$, $V_{\mathbb{Z}}(x^2 + 1) = V_{\mathbb{R}}(x^2 + 1) = \emptyset$ while $V_{\mathbb{C}}(x^2 + 1) = \{i, -i\}$. Thus the variety of a polynomial in $R[\mathbf{x}]$ depends on the coefficient ring R.

If f is a non-zero constant — i.e., a non-zero element of R — then $V_R(f) = \emptyset$. This is true no matter what R is. What conditions on R do we need to ensure that all polynomials in $R[\mathbf{x}]$ other than the constant polynomials have non-empty varieties? To investigate an example, let us start with $R = \mathbb{Z}$ and consider the **univariate** polynomial ring $\mathbb{Z}[x]$. Then we see that $V_{\mathbb{Z}}(x^2 - 2) = \emptyset$ since $\sqrt{2} \in \mathbb{R}$ but not in $\mathbb{Z}$. So we enlarge our coefficient ring from $\mathbb{Z}$ to $\mathbb{R}$. However, $V_{\mathbb{R}}(x^2 + 1) = \emptyset$ while $V_{\mathbb{C}}(x^2 + 1) = \{i, -i\}$. This shows that we need to enlarge R to $\mathbb{C}$. In fact $\mathbb{C}$ is good enough. The **fundamental theorem of algebra** states that every non-constant polynomial in $\mathbb{C}[x]$ of degree d has exactly d zeros counting multiplicities.

Definition 10.5. A field F is **algebraically closed** if every polynomial in the univariate polynomial ring $F[x]$ has a root in F.

It requires work to show that algebraically closed fields exist and that every field is contained in an algebraically closed field called its **algebraic closure**. For instance, the algebraic closure of $\mathbb{R}$ and $\mathbb{Q}$ is $\mathbb{C}$. From now on we will mostly be interested in polynomial rings over algebraically closed fields. In fact, we mostly consider the polynomial ring $\mathbb{C}[x_1, \ldots, x_n]$ although occasionally we will also consider polynomial rings over $\mathbb{Z}, \mathbb{Q}$ and $\mathbb{R}$.

Exercise 10.6. Check that $V_{\mathbb{R}}(x^2 + y^2 - 1)$ is the unit circle in $\mathbb{R}^2$ centered at the origin. What do you expect $V_{\mathbb{C}}(x^2 + y^2 - 1)$ to look like? What is $V_{\mathbb{Z}}(x^2 + y^2 - 1)$?

An **algebraic set** in R^n is any subset of R^n that consists of the zeros of a system of polynomials — not necessarily finite. Clearly, $V_R(f)$ is an algebraic set for all $f \in R[\mathbf{x}]$. A variety cut out by a single non-zero polynomial is called a **hypersurface**. We now examine varieties cut out by more than one polynomial. It is most efficient to do this via *polynomial ideals*. From now on we always consider polynomial rings over algebraically closed fields $\mathbf{k}$.

Definition 10.7. A subset $I \subseteq \mathbf{k}[\mathbf{x}]$ is an **ideal** if (1) $(I, +)$ is an abelian subgroup of $\mathbf{k}[\mathbf{x}]$ and (2) for all $h \in \mathbf{k}[\mathbf{x}]$ and $g \in I$, $hg \in I$.

Note that by property (2), $I = \mathbf{k}[\mathbf{x}]$ if any non-zero element c of $\mathbf{k}$ is in I. This is because if $c \in I$, then $c^{-1}c = 1 \in I$ by property (2) where we take $c^{-1} \in \mathbf{k} \subset \mathbf{k}[\mathbf{x}]$ and $c \in I$. This in turn implies that for any $f \in k[\mathbf{x}]$, $f \cdot 1 = f \in I$ again by property (2).

Definition 10.8. Given a set of polynomials $P \subset \mathbf{k}[\mathbf{x}]$, the ideal *generated* by P is the set

$$\langle P \rangle := \{\sum h_p p \ : \ h_p \in \mathbf{k}[\mathbf{x}], \ p \in P\}.$$

We say that P is a **basis** of the ideal $\langle P \rangle$.

In the 1890's Hilbert proved that every polynomial ideal in $\mathbf{k}[\mathbf{x}]$ has a finite basis. This theorem is known as *Hilbert's basis theorem* and is a cornerstone of commutative algebra and algebraic geometry. We will see a proof of this theorem in Chapter 12. Let's take a look at some examples of ideals.

Example 10.9.   (1) A **principal ideal** is generated by a single polynomial $f \in \mathbf{k}[\mathbf{x}]$. By definition, $\langle f \rangle = \{hf : h \in \mathbf{k}[\mathbf{x}]\}$. All elements of $\langle f \rangle$ are divisible by the generator f.

(2) Every matrix A with entries in $\mathbf{k}$ defines an ideal as follows. If $\mathbf{a} = (a_1, \ldots, a_n)$ is a row of A, then define $f_{\mathbf{a}} := a_1 x_1 + a_2 x_2 + \cdots + a_n x_n \in \mathbf{k}[x_1, \ldots, x_n]$. The ideal generated by the linear polynomials $f_{\mathbf{a}}$ as $\mathbf{a}$ runs over the rows of A is called the **linear ideal** of the matrix A. Note that this ideal contains non-linear polynomials such as $f_{\mathbf{a}}^2$.

(3) If all the generators of an ideal are monomials, we have a **monomial ideal**. Note that there are many non-monomials in a monomial ideal. However, if a polynomial $f = \sum c_{\mathbf{a}} \mathbf{x}^{\mathbf{a}}$ is in a monomial ideal, then every $\mathbf{x}^{\mathbf{a}}$ is also in the monomial ideal. This implies that $\mathbf{x}^{\mathbf{a}}$ is divisible by a (monomial) generator of the ideal. This observation allows one to "draw" monomial ideals as follows. Identify a monomial $\mathbf{x}^{\mathbf{a}}$ in the monomial ideal M with its exponent vector $\mathbf{a} \in \mathbb{N}^n$. Draw M as the collection $\{\mathbf{a} : \mathbf{x}^{\mathbf{a}} \in M\}$. This collection of lattice points in $\mathbb{N}^n$ is in bijection with the monomials in M.

Exercise 10.10.   (1) Draw the monomial ideal $\langle x^3, x^2 y^2, xy^4 \rangle$.

(2) If $M = \langle \mathbf{x}^{\mathbf{m}_1}, \cdots, \mathbf{x}^{\mathbf{m}_t} \rangle$, then check that the picture of M is precisely

$$\bigcup_{i=1}^{t} (\mathbf{m}_i + \mathbb{N}^n).$$

(3) Let $M = \langle \mathbf{x}^{\alpha} : \alpha \in \mathcal{I} \rangle$. Then show that $\mathbf{x}^{\beta} \in M$ if and only if $\mathbf{x}^{\beta}$ is divisible by $\mathbf{x}^{\alpha}$ for some $\alpha \in \mathcal{I}$.

(4) Two monomial ideals are the same if and only if they contain the same monomials.

Exercise 10.11. Prove Hilbert's basis theorem for monomial ideals. (You need to show that if an ideal is generated by a collection of monomials (possibly infinite), then the ideal is in fact generated by a finite subset of the above basis.)

Hints (From [**CLO92**]): Let $M = \langle \mathbf{x}^\alpha : \alpha \in \mathcal{I} \rangle \subset \mathbf{k}[\mathbf{x}]$ be a monomial ideal where $\mathcal{I}$ is some, possibly infinite, index set. We want to show that there exists $\alpha(1), \ldots, \alpha(s) \in \mathcal{I}$ such that $M = \langle \mathbf{x}^{\alpha(1)}, \ldots, \mathbf{x}^{\alpha(s)} \rangle$.

(1) Approach this problem by induction on the number of variables in $\mathbf{k}[x_1, \ldots, x_n]$.

(2) Write each monomial in $\mathbf{k}[x_1, \ldots, x_n]$ as $\mathbf{x}^\alpha x_n^m$ where $\alpha \in \mathbb{N}^{n-1}$. Let J be the ideal in $\mathbf{k}[x_1, \ldots, x_{n-1}]$ defined by $\langle \mathbf{x}^\alpha : \mathbf{x}^\alpha x_n^m \in M \rangle$. With (1) in mind, what can you say about J?

(3) Check that J can be generated by finitely many elements, say

$$\{\mathbf{x}^{\alpha(1)}, \ldots, \mathbf{x}^{\alpha(t)}\}$$

with $\mathbf{x}^{\alpha(i)} x_n^{m_i} \in M$ for each $1 \leq i \leq t$. Let

$$m := \max\{m_1, \ldots, m_t\}$$

and for every k, $0 \leq k \leq m - 1$, consider the ideal $J_k \subset \mathbf{k}[x_1, \ldots, x_{n-1}]$ generated by monomials $\mathbf{x}^\beta$ such that $\mathbf{x}^\beta x_n^k \in M$. Each J_k has a finite generating set $\{\mathbf{x}^{\alpha_k(1)}, \ldots, \mathbf{x}^{\alpha_k(t_k)}\}$.

(4) Define $\mathcal{S} = \{\mathbf{x}^{\alpha(1)} x_n^m, \ldots, \mathbf{x}^{\alpha(t)} x_n^m\}$ and $\mathcal{S}_k = \{\mathbf{x}^{\alpha_k(1)} x_n^k, \ldots, \mathbf{x}^{\alpha_k(t_k)} x_n^k\}$ for every $0 \leq k \leq m - 1$. Show that $\mathcal{S} \cup \mathcal{S}_0 \cup \cdots \cup \mathcal{S}_{m-1}$ is a generating set for M.

Definition 10.12. The **variety** of a collection of polynomials $P \subseteq \mathbf{k}[\mathbf{x}]$ is the set

$$V_{\mathbf{k}}(P) := \{(r_1, \ldots, r_n) \in \mathbf{k}^n : p(r_1, \ldots, r_n) = 0 \text{ for all } p \in P\}.$$

Exercise 10.13. (1) Show that

$$V_{\mathbf{k}}(\{f_1, \ldots, f_t\}) = V_{\mathbf{k}}(\langle f_1, \ldots, f_t \rangle).$$

Then, by Hilbert's basis theorem, the variety of any polynomial ideal is also the zero-set of a finite number of polynomials.

(2) Compute $V_{\mathbb{R}}(x^2 + y^2 - 1, x - 1/2)$. Using (1), find a different set of polynomials that cut out the same variety.

(3) Compute $V_{\mathbb{R}}(\langle x^3, x^2y^2, xy^4 \rangle)$ and compare it to $V_{\mathbb{R}}(\langle x \rangle)$ where $\langle x \rangle \subset \mathbb{R}[x, y]$.

(4) What is the general form of the variety (over $\mathbb{R}$) of a monomial ideal $M = \langle \mathbf{x^{m_1}}, \cdots, \mathbf{x^{m_t}} \rangle$? Does it matter what the field $\mathbf{k}$ is?

Definition 10.14. The **radical** of an ideal $I \subseteq \mathbf{k}[\mathbf{x}]$ is the ideal $\sqrt{I} := \{f \in \mathbf{k}[\mathbf{x}] : f^r \in I \text{ for some power } r \in \mathbb{N}\}$.

Check that $\sqrt{I}$ is an ideal and that it contains I. We say that I is a **radical ideal** if $I = \sqrt{I}$. For a monomial $\mathbf{x^m} \in \mathbf{k}[\mathbf{x}]$ define its support to be $\mathrm{supp}(\mathbf{x^m}) := \{i : m_i > 0\}$. Be careful not to be confused by the use of "support" in two different ways. The support of a polynomial is a collection of vectors. In this sense, the support of a monomial should be the set containing its exponent vector. However, what one usually means by the support of a monomial $\mathbf{x^m}$ is the set of indices $\{i : m_i > 0\}$ which coincides with, $\mathrm{supp}(\mathbf{m})$, the support of the vector $\mathbf{m}$.

Exercise 10.15.   (1) Show that $\sqrt{\langle x^3, x^2y^2, xy^4 \rangle} = \langle x \rangle$.

(2) In general, if $M = \langle \mathbf{x^{m_1}}, \cdots, \mathbf{x^{m_t}} \rangle$ is a monomial ideal, then show that

$$\sqrt{M} = \langle \prod_{j \in \mathrm{supp}(\mathbf{x^{m_i}})} x_j, \ i = 1 \ldots, t \rangle.$$

(3) Furthermore, show that both M and $\sqrt{M}$ have the same variety over any field $\mathbf{k}$. Similarly, I and $\sqrt{I}$ have the same variety over any field $\mathbf{k}$.

This brings us to a second famous theorem of Hilbert from the 1890's.

Theorem 10.16. Hilbert's Nullstellensatz:
Weak form: *Let $\mathbf{k}$ be an algebraically closed field and let $I \subset \mathbf{k}[\mathbf{x}]$ be an ideal. Then $V_{\mathbf{k}}(I) = \emptyset$ if and only if $I = \mathbf{k}[\mathbf{x}]$.*
Strong form: *Let $\mathbf{k}$ be an algebraically closed field. If $I \subset \mathbf{k}[\mathbf{x}]$ is an*

ideal, then the set of all polynomials that vanish on $V_{\mathbf{k}}(I)$ is precisely the polynomials in $\sqrt{I}$.

The weak Nullstellensatz says that a system of polynomials in $\mathbf{k}[\mathbf{x}]$ has at least one solution if and only if the ideal generated by the polynomials is not all of $\mathbf{k}[\mathbf{x}]$. The strong Nullstellensatz shows that many different ideals in $\mathbf{k}[\mathbf{x}]$ can cut out the same variety in $\mathbf{k}^n$. A simple example to keep in mind is the ideals $\langle x^l \rangle$ in $\mathbf{k}[x]$ for all $l = 1, 2, \ldots$ which cut out the same variety $\{0\} \subset \mathbf{k}$. However, there is a one-to-one correspondence between radical ideals in $\mathbf{k}[\mathbf{x}]$ and varieties in $\mathbf{k}^n$. This creates an algebra-geometry dictionary that allows us to pass from geometric objects such as varieties to algebraic objects such as ideals which are easier to manipulate. See Chapter 4 in [**CLO92**] for more details.

Chapter 11

Gröbner Bases I

This chapter and the next aim to give a brief introduction to the basics of Gröbner basis theory. By now, there are many excellent books on Gröbner bases and their applications such as [**AL94**], [**CLO92**], [**CLO98**], [**GP02**] and [**KR00**]. Our account here will be brief.

From now on, let $S := \mathbf{k}[x_1, \dots, x_n] = \mathbf{k}[\mathbf{x}]$ be the polynomial ring in n variables over an algebraically closed field $\mathbf{k}$. Let I be an ideal of S. By Hilbert's basis theorem I has a finite generating set. Gröbner bases of I are special finite generating sets of I. We first motivate the need for these bases and then define and construct them.

Given an ideal $I \subset S$ and a polynomial $f \in S$, a fundamental problem is to decide whether f belongs to I. This is known as the *ideal membership problem*. We will see shortly that Gröbner bases can be used to solve this problem. We begin by examining algorithms for ideal membership in two familiar families of ideals.

(i) **Univariate ideals**: (See [**CLO92**, Chapter 1].) Consider the univariate polynomial ring $\mathbf{k}[x]$. If $\deg(f)$ denotes the degree of a polynomial $f \in S$, then f has the form

$$f = k_p x^p + k_{p-1} x^{p-1} + \cdots + k_0$$

where $p = \deg(f)$, $k_p \neq 0$, and all $k_i \in \mathbf{k}$. We say that f is **monic** if its **leading coefficient** k_p equals one. Given two polynomials f

and g in S with $\deg(f) \geq \deg(g)$, we can divide f by g to get unique polynomials h and $r \in \mathbf{k}[x]$ such that $f = hg + r$ where r is the remainder and $\deg(r) < \deg(g)$. This is the usual long division of f by g that you may remember from high school.

Exercise 11.1. (1) Divide $x^4 + 2x + 3$ by $5x^2 + 2$.
(2) Write a pseudo-code for the division algorithm in $\mathbf{k}[x]$.

Theorem 11.2. *Every ideal of* $\mathbf{k}[x]$ *is principal. A non-zero ideal* I *in* $\mathbf{k}[x]$ *is generated by any non-zero polynomial in it of lowest degree.*

Exercise 11.3. Let us outline a proof of Theorem 11.2. Let f be a non-zero polynomial in I of lowest degree. Then clearly $\langle f \rangle \subseteq I$. So the work is to show that if $g \in I$, then $g \in \langle f \rangle$ or, equivalently, that f divides g. Since $\deg(g) \geq \deg(f)$, we can divide g by f to get the unique polynomials q and r such that $g = qf + r$ with $\deg(r) < \deg(f)$. If $r = 0$, then $g \in \langle f \rangle$. If $r \neq 0$, can you derive a contradiction, and thus conclude that $r = 0$ as we want?

Definition 11.4. The g.c.d. of two polynomials f and g in $\mathbf{k}[x]$ is a polynomial h such that (1) h divides both f and g and (2) if p is a polynomial that divides f and g, then p divides h.

Given two polynomials f and g in $\mathbf{k}[x]$, we can compute their g.c.d. by the **Euclidean algorithm** which works as follows. See [**CLO92**, Chapter 1] for a proof of its correctness. We use $\mathrm{rem}(h, s)$ to denote the remainder of h on division by s.

Input: f, g
Initialize: $h := f, s := g$
While $s \neq 0$ **do**
   $r := \mathrm{rem}(h, s)$, $h := s$, $s := r$
Output: $h = \mathrm{g.c.d.}(f, g)$

Theorem 11.5. ([**CLO92**, Proposition 6, §1.5])

(1) If $f, g \in \mathbf{k}[x]$*, then* $\mathrm{g.c.d.}(f, g)$ *exists and is unique up to multiplication by a non-zero constant.*

(2) The ideal $\langle f, g \rangle$ *is generated by* $\mathrm{g.c.d.}(f, g)$*.*

Exercise 11.6. Prove part (2) of Theorem 11.5.

G.c.d.s can be computed for more than two polynomials in $\mathbf{k}[x]$.

Definition 11.7. The g.c.d. of polynomials $f_1, \ldots, f_s \in \mathbf{k}[x]$ is a polynomial h such that (1) h divides $f_1, \ldots, f_s$ and (2) if p is a polynomial that divides $f_1, \ldots, f_s$, then p divides h.

Theorem 11.8. ([**CLO92**, Proposition 8, §1.5])

    (1) The g.c.d. of $f_1, \ldots, f_s$ exists and is unique up to multiplication by a non-zero constant.

    (2) The ideal $\langle f_1, \ldots, f_s \rangle$ is generated by g.c.d.$(f_1, \ldots, f_s)$.

    (3) If $s \geq 3$, then g.c.d.$(f_1, \ldots, f_s) =$ g.c.d.$(f_1, \text{g.c.d.}(f_2, \ldots, f_s))$.

Exercise 11.9. Show that the variety of the ideal $\langle x^2 + 1, x^3 - 2x + 1 \rangle$ in $\mathbb{C}[x]$ is empty.

We now have algorithms for solving the following three fundamental problems for a univariate polynomial ideal $I \subseteq \mathbf{k}[x]$.

- **Finding a basis for** I: If $I = \langle f_1, \ldots, f_t \rangle \subset \mathbf{k}[x]$, then I is generated by $g = $ g.c.d.$(f_1, \ldots, f_t)$, which can be computed by the Euclidean algorithm.

- **Ideal membership**: If $f \in \mathbf{k}[x]$, then $f \in I = \langle g \rangle$ if and only if the remainder obtained by dividing f by g is zero. Thus ideal membership can be determined by the division algorithm.

- **Solving** $\{f_1 = f_2 = \cdots = f_t = 0\}$, $f_i \in \mathbf{k}[x]$: The variety $V_{\mathbf{k}}(f_1, \ldots, f_t) = V_{\mathbf{k}}(g)$ where $g = $ g.c.d.$(f_1, \ldots, f_t)$. The roots of g can be found via radicals when its degree is small and by numerical methods otherwise.

(ii) **Linear ideals**:([**Stu96**, Chapter 1]) Let $A \in \mathbb{Z}^{d \times n}$ be a matrix of rank d. Recall that the linear ideal of A is

$$I = \left\langle \sum_{j=1}^{n} a_{ij} x_j \; : \; i = 1, \ldots, d \right\rangle \subset \mathbb{R}[\mathbf{x}].$$

The variety $V_{\mathbb{R}}(I)$ is the $(n-d)$-dimensional vector space

$$\ker_{\mathbb{R}}(A) = \{\mathbf{p} \in \mathbb{R}^n \; : \; A\mathbf{p} = \mathbf{0}\}.$$

A non-zero linear form f in I is a **circuit** of I if f has minimal support (with respect to inclusion) among all polynomials in I. The coefficient vector of a circuit of I is therefore a vector in the row span of A of minimal support. Let B be any $(n - d) \times n$ integer matrix whose rows form a basis of $\ker_{\mathbb{R}}(A)$. Then the vectors in the row span of A are precisely the linear dependencies on the columns of B. The coefficient vectors of the circuits of I are therefore the dependencies on the columns of B of minimal support.

For $J \subseteq [n]$ with $|J| = d$, let $\det(A_J)$ be the determinant of A_J, the submatrix of A with column indices J. The following algorithm computes the circuits of I.

Algorithm 11.10. ([**Stu02**, Chapter 8]) For any $(n - d + 1)$-subset $\tau = \{\tau_1, \ldots, \tau_{n-d+1}\} \subseteq [n]$, form the vector

$$C_\tau := \sum_{i=1}^{n-d+1} (-1)^i \cdot \det(B_{\tau \setminus \{\tau_i\}}) \cdot \mathbf{e}_{\tau_i}$$

where $\mathbf{e}_j$ is the jth unit vector of $\mathbb{R}^n$. If C_τ is non-zero, then compute the primitive vector obtained by dividing through with the g.c.d. of its components. The resulting vector is a circuit and all circuits are obtained this way.

Example 11.11. Let $A = \begin{pmatrix} 1 & 2 & 3 & 4 & 5 \\ 6 & 7 & 8 & 9 & 10 \end{pmatrix}$. Then

$$I = \langle x_1 + 2x_2 + 3x_3 + 4x_4 + 5x_5, \ 6x_1 + 7x_2 + 8x_3 + 9x_4 + 10x_5 \rangle.$$

The rows of

$$B = \begin{pmatrix} 3 & -4 & 0 & 0 & 1 \\ 2 & -3 & 0 & 1 & 0 \\ 1 & -2 & 1 & 0 & 0 \end{pmatrix}$$

form a basis for $\ker_{\mathbb{R}}(A)$. Let us compute one of the circuits of A. For $\tau = \{1, 2, 3, 4\}$,

$$C_\tau = -\det \begin{pmatrix} -4 & 0 & 0 \\ -3 & 0 & 1 \\ -2 & 1 & 0 \end{pmatrix} \mathbf{e}_1 + \det \begin{pmatrix} 3 & 0 & 0 \\ 2 & 0 & 1 \\ 1 & 1 & 0 \end{pmatrix} \mathbf{e}_2$$

$$-\det \begin{pmatrix} 3 & -4 & 0 \\ 2 & -3 & 1 \\ 1 & -2 & 0 \end{pmatrix} \mathbf{e}_3 + \det \begin{pmatrix} 3 & -4 & 0 \\ 2 & -3 & 0 \\ 1 & -2 & 1 \end{pmatrix} \mathbf{e}_4$$

which equals $(-4, -3, -2, -1, 0)$. Hence $4x_1 + 3x_2 + 2x_3 + x_4$ is a circuit of I.

Exercise 11.12. Compute all circuits of the linear ideal of the matrix A in the above example. What is a natural upper bound for the number of circuits of the linear ideal of a $d \times n$ matrix A?

Proposition 11.13. *Let $C = (c_{ij}) \in \mathbb{R}^{d \times n}$ be the Gauss-Jordan form (reduced row-echelon form) of A and let $g_i = \sum_{j=1}^{n} c_{ij}x_j$ be the linear forms corresponding to the rows of C. Let I be the linear ideal of A. Then*

(1) $\{g_1, \ldots, g_d\}$ is a minimal generating set for I and

(2) the linear forms $g_1, \ldots, g_d$ are circuits of I.

Proof.   (1) Since every row of C is a linear combination of rows of A and vice versa, every g_i is a linear combination of the f_i's and every f_i a linear combination of the g_i's. Thus $I = \langle f_1, \ldots, f_d \rangle = \langle g_1, \ldots, g_d \rangle$. Since C is in reduced row-echelon form, we may assume that $C = [I \,|\, E]$ where I is the $d \times d$ identity matrix. Therefore, $g_i = x_i + \sum_{j=d+1}^{n} e_{ij}x_j$ for each $i = 1, \ldots, d$. This implies that no g_i is a linear combination of $\{g_j, j \neq i\}$ and thus $\{g_1, \ldots, g_d\}$ is a minimal generating set for I.

(2) Again assume without loss of generality that $g_i = x_i + \sum_{j=d+1}^{n} e_{ij}x_j$ for each $i = 1, \ldots, d$. Suppose g_1 is not a circuit. Then there exists a linear polynomial $g \in I$ such that $\operatorname{supp}(g) \subsetneq \operatorname{supp}(g_1)$. However, $g = t_1g_1 + \ldots + t_dg_d$ for scalars $t_1, \ldots, t_d \in \mathbb{R}$. Since $\operatorname{supp}(g) \subset \operatorname{supp}(g_1)$, $t_2 = t_3 = \cdots = t_d = 0$. This implies that $g = t_1g_1$, $t_1 \neq 0$, and hence $\operatorname{supp}(g) = \operatorname{supp}(g_1)$, a contradiction. The same argument can be repeated for $g_2, \ldots, g_d$.

$\square$

Proposition 11.14. *Assume the notation as in Proposition 11.13 and its proof.*

(1) A polynomial $f \in S$ lies in I if and only if successively replacing every occurrence of x_i, $i = 1, \ldots, d$, in f, with $-\sum_{j=d+1}^{n} e_{ij}x_j$ results in the zero polynomial.

(2) *To solve the linear system* $A\mathbf{x} = \mathbf{0}$, *we back solve the "tri-angularized" system* $g_1 = g_2 = \cdots = g_d = 0$.

Proof.   (1) Let f' be obtained from f by successively replacing every x_i, $i = 1, \ldots, d$, in f, with $-\sum_{j=d+1}^{n} e_{ij} x_j$. Then $f' \in \mathbb{R}[x_{d+1}, \ldots, x_n]$. This implies that $f = \sum_{i=1}^{d} h_i g_i + f'$ where $h_i \in \mathbb{R}[\mathbf{x}]$. If $f' = 0$, then clearly $f \in I$. Conversely, if $f \in I$, then $f' = f - \sum_{i=1}^{d} h_i g_i \in I \cap \mathbb{R}[x_{d+1}, \ldots, x_n] = \{0\}$. The last equality follows from the fact that no polynomial combination of $g_1, \ldots, g_d$ can lie in $\mathbb{R}[x_{d+1}, \ldots, x_n]$ because of the special structure of the g_i's.

(2) This is the familiar method from linear algebra of solving linear systems by Gaussian elimination.

$\square$

Proposition 11.14(1) suggests a division algorithm for linear polynomials in many variables that succeeds in determining ideal membership for linear ideals. Note that when we perform Gauss-Jordan elimination on A to obtain $C = [I \mid E]$, we are implicitly ordering the variables in S such that $x_1 > x_2 > \cdots > x_n$. The division algorithm suggested in Proposition 11.14(1) replaces every occurrence of the "leading term" x_i in g_i with $x_i - g_i$, which is the sum of the "trailing terms" in g_i.

Thus the questions we started with have well-known algorithms and answers when I is either a univariate principal ideal or a multivariate linear ideal. We now seek a common generalization of these methods to multivariate polynomials of arbitrary degrees and their ideals. This will lead us to Gröbner bases of polynomial ideals.

In order to mimic the procedures we saw thus far, we first need to impose an ordering on the monomials in $S = \mathbf{k}[\mathbf{x}]$ so that the terms in a polynomial are always ordered. This is important if the generalized division algorithm is to replace the leading term of a divisor by the sum of its trailing terms. A **total order** on a set T is an ordering of the elements of T in which any two elements are comparable. For example, the usual "$\leq$" ordering of the integers is a total order on $\mathbb{Z}$. A **partial order** on T is an ordering where not all pairs of elements may be comparable. For example, the vectors $(2,3,0)$ and $(1,4,0)$ are incomparable if we order $\mathbb{Z}^3$ by the usual "$\leq$" order which makes $(a_1, a_2, a_3) \leq (b_1, b_2, b_3)$ if and only if $a_1 \leq b_1, a_2 \leq b_2$ and $a_3 \leq b_3$.

Definition 11.15. A **term order** $\succ$ on S is a total order on the monomials of S such that

(1) $\mathbf{x^a} \succ \mathbf{x^b}$ implies that $\mathbf{x^a x^c} \succ \mathbf{x^b x^c}$ for all $\mathbf{c} \in \mathbb{N}^n$ and

(2) $\mathbf{x^a} \succ \mathbf{x^0} = 1$ for all $\mathbf{a} \in \mathbb{N}^n$.

Example 11.16. The most common examples of term orders are the *lexicographic* and the *reverse lexicographic* orders on S with respect to a fixed ordering of the variables. Suppose $x_1 \succ x_2 \succ \cdots \succ x_n$.

In the **lexicographic order**, $\mathbf{x^a} \succ \mathbf{x^b}$ if and only if the left-most non-zero term in $\mathbf{a} - \mathbf{b}$ is positive. For example, if $x \succ y \succ z$, then

$$x^3 \succ x^2 y \succ y^{100} \succ y^{99} z^{10000} \succ y^{99}.$$

In the (graded) **reverse lexicographic order**, $\mathbf{x^a} \succ \mathbf{x^b}$ if and only if either $\deg(\mathbf{x^a}) > \deg(\mathbf{x^b})$, or $\deg(\mathbf{x^a}) = \deg(\mathbf{x^b})$ and the right-most non-zero term in $\mathbf{a} - \mathbf{b}$ is negative. The degree of a monomial is its usual total degree. Again if $x \succ y \succ z$, then

$$x^3 \succ x^2 y \succ xy^2 \succ y^3 \succ x^2 z \succ xyz \succ y^2 z \succ xz^2 \succ yz^2 \succ z^3.$$

The reverse lexicographic order defined here is *degree-compatible*, which means that it first compares two monomials by degree and then breaks ties using the rule described. Note that there are $n!$ lex (lexicographic) and revlex (graded reverse lexicographic) orders in S.

Exercise 11.17. Consider the following term orders on $\mathbf{k}[x, y]$:
(i) lex ordering with $x \succ y$ and
(ii) graded reverse lex ordering with $x \succ y$.
Recall that we can identify the monomials in $\mathbf{k}[x, y]$ by the elements of $\mathbb{N}^2$. Draw two copies of $\mathbb{N}^2$ and see how the lattice points in each copy get ordered by (i) and (ii).

Exercise 11.18. Show that the revlex order is not a term order if we do not require it to be degree compatible — i.e., if we define it as $\mathbf{x^a} \succ \mathbf{x^b}$ if and only if the right-most non-zero term in $\mathbf{a} - \mathbf{b}$ is negative.

Exercise 11.19. How many distinct term orders does $\mathbf{k}[x]$ have?

There are many other examples of term orders. In fact, every vector in $\mathbb{R}^n_{\geq 0}$ can be used to define a term order as follows. Given $\omega \in \mathbb{R}^n_{\geq 0}$, pick a fixed term order $\succ$ on S such as a revlex order, and define the total order $\succ_\omega$ as follows:

$$\mathbf{x^a} \succ_\omega \mathbf{x^b} \text{ if either } \omega \cdot \mathbf{a} > \omega \cdot \mathbf{b} \text{ or } \omega \cdot \mathbf{a} = \omega \cdot \mathbf{b} \text{ and } \mathbf{x^a} \succ \mathbf{x^b}.$$

All term orders can be mimicked by weight vectors when the polynomials we are dealing with have bounded degrees. For instance, the lex order on $\mathbf{k}[x, y]$ with $x \succ y$ can be mimicked by the weight vector $\omega = (100, 1)$ as long as the degrees of all monomials we will see are less than 100. In general, we can use the weight vector $(100, 100\epsilon)$, where ϵ can be made as small as we want depending on the largest degree of a monomial that we will encounter. In S we could mimic the lex order with $x_1 \succ \cdots \succ x_n$ using

$$\omega = (100, 100\epsilon, 100\epsilon^2, \ldots, 100\epsilon^{n-1}).$$

Exercise 11.20. What weight vector can be used to mimic a revlex order?

We now fix a term order $\succ$ on S which immediately orders the terms in a polynomial. The unique term in a polynomial f that contains the highest monomial with respect to $\succ$ is called the **leading term** or **initial term** of f. We denote this term by $\mathrm{LT}_\succ(f)$ and write $f = \mathrm{LT}_\succ(f) + f'$. Using this, we can attempt to write down a division algorithm for multivariate polynomials. The following algorithm is copied verbatim from [**CLO92**, §2.3, Theorem 3].

Algorithm 11.21. Division algorithm for multivariate polynomials ([**CLO92**, Theorem 3]).
INPUT: A dividend $f \in S$ and an ordered set of divisors $F = [f_1, \ldots, f_s]$ where each f_i lies in S.
OUTPUT: Polynomials $a_1, \ldots, a_s, r \in S$, such that $f = \sum_{i=1}^s a_i f_i + r$ where either $r = 0$ or no term in r is divisible by $\mathrm{LT}_\succ(f_1), \ldots, \mathrm{LT}_\succ(f_s)$.
INITIALIZE: $a_1 := 0, \ldots, a_s := 0, r := 0; \ p := f$
WHILE $p \neq 0$ DO
    $i := 1$

divisionoccurred := false
WHILE $i \leq s$ AND divisionoccurred = false DO
    IF $\mathrm{LT}_\succ(f_i)$ divides $\mathrm{LT}_\succ(p)$ THEN
        $a_i := a_i + \mathrm{LT}_\succ(p)/\mathrm{LT}_\succ(f_i)$
        $p := p - (\mathrm{LT}_\succ(p)/\mathrm{LT}_\succ(f_i))f_i$
        divisionoccurred := true
    ELSE
        $i := i + 1$
    IF divisionoccurred = false THEN
        $r := r + \mathrm{LT}_\succ(p)$
        $p := p - \mathrm{LT}_\succ(p)$

Exercise 11.22. ([**CLO92**], Chapter 2.3, Examples 2 and 4])

Fix the lex order on $\mathbf{k}[x, y]$ such that $x \succ y$.

(1) Divide $f = x^2 y + x y^2 + y^2$ by the ordered list of polynomials $[f_1 = xy - 1, f_2 = y^2 - 1]$ and record the remainder.

(2) Now repeat the division with the list of divisors reordered as $[f_2 = y^2 - 1, f_1 = xy - 1]$. Record the remainder.

Note that the remainders are different. This example shows that the above division algorithm for multivariate polynomials has several drawbacks, one of which is that it does not produce unique remainders. This makes it impossible to check ideal membership of f in $\langle f_1, f_2 \rangle$ by dividing f with the generators f_1, f_2.

The above example shows that arbitrary generating sets of ideals and a naive extension of the usual division algorithm cannot be used for ideal membership. We will see that this difficulty disappears when the basis of the ideal is a Gröbner basis.

Chapter 12

Gröbner Bases II

Given a term order $\succ$ on $S = \mathbf{k}[\mathbf{x}]$ and a polynomial $f = \sum m_{\mathbf{a}}\mathbf{x}^{\mathbf{a}}$, recall that the leading term or initial term of f is the term $m_{\mathbf{a}}\mathbf{x}^{\mathbf{a}}$ in f such that $\mathbf{x}^{\mathbf{a}} \succ \mathbf{x}^{\mathbf{a}'}$ for all $\mathbf{a}' \in \mathrm{supp}(f)$ different from $\mathbf{a}$. It is denoted as $\mathrm{LT}_{\succ}(f)$. The **leading monomial** or **initial monomial** of f is the monomial $\mathbf{x}^{\mathbf{a}}$ in $\mathrm{LT}_{\succ}(f)$. It is denoted as $\mathrm{in}_{\succ}(f)$.

Example 12.1. Let $f = 3x_1 x_3^2 + \sqrt{2}x_3^2 - x_1 x_2^2 \in \mathbb{C}[x_1, x_2, x_3]$ and let $\succ$ be the reverse lexicographic order with $x_1 \succ x_2 \succ x_3$. Then $\mathrm{in}_{\succ}(f) = x_1 x_2^2$ is the initial monomial of f and $-x_1 x_2^2$ is the initial term of f.

The **initial ideal** of an ideal $I \subset S$ with respect to $\succ$ is the monomial ideal:

$$\mathrm{in}_{\succ}(I) := \langle \mathrm{in}_{\succ}(f) : f \in I \rangle \subseteq S.$$

The monomials of S that do not lie in a monomial ideal M are called the **standard monomials** of M. Recall that M can be depicted by its staircase diagram in $\mathbb{N}^n$, which is the collection of all exponent vectors of monomials in M. Clearly, this set of "dots" is closed under the addition of $\mathbb{N}^n$ to any of the dots. Equivalently, its complement in $\mathbb{N}^n$ is a **down-set** or **order ideal** in $\mathbb{N}^n$. This means that if $\mathbf{x}^{\mathbf{u}} \notin M$, then $\mathbf{x}^{\mathbf{v}} \notin M$ for all $\mathbf{v} \leq \mathbf{u}$, where $\leq$ is the usual componentwise partial order on $\mathbb{N}^n$.

Example 12.2. Let $I = \langle x^2 - y, x^3 - x \rangle \subset \mathbf{k}[x, y]$ and let $\succ$ be the lexicographic order with $x \succ y$. The polynomial $x(x^2 - y) - (x^3 - x) = -xy + x \in I$, which shows that $\mathrm{in}_\succ(I) \supset \langle x^2, x^3, xy \rangle = \langle x^2, xy \rangle$. We will see later how to compute all of $\mathrm{in}_\succ(I)$.

By the Hilbert basis theorem for monomial ideals (Exercise 10.11) all monomial ideals of S have a minimal finite generating set consisting of monomials. It can also be seen quite easily that this minimal generating set is unique. Suppose $\{\mathbf{x}^{\mathbf{m}_1}, \ldots, \mathbf{x}^{\mathbf{m}_s}\}$ is the unique minimal finite generating set of $\mathrm{in}_\succ(I)$. Then by definition, there exists $g_1, \ldots, g_s \in I$ such that $\mathrm{in}_\succ(g_i) = \mathbf{x}^{\mathbf{m}_i}$, $i = 1, \ldots, s$. We also have that $\mathrm{in}_\succ(I) = \langle \mathrm{in}_\succ(g_1), \ldots, \mathrm{in}_\succ(g_s) \rangle$. We call $\mathcal{G}_\succ(I) := \{g_1, \ldots, g_s\}$ a *Gröbner basis* of I with respect to $\succ$.

Definition 12.3.   (1) A finite set of polynomials $\{g_1, \ldots, g_s\} \subset I$ is a **Gröbner basis** of I with respect to $\succ$ if $\mathrm{in}_\succ(I) = \langle \mathrm{in}_\succ(g_1), \ldots, \mathrm{in}_\succ(g_s) \rangle$. We assume that each $\mathrm{in}_\succ(g_i)$ is a monomial (with coefficient one).

   (2) Furthermore, if $\{\mathrm{in}_\succ(g_1), \ldots, \mathrm{in}_\succ(g_s)\}$ is the unique minimal generating set of $\mathrm{in}_\succ(I)$, we say that $\mathcal{G}_\succ(I)$ is a **minimal** Gröbner basis of I with respect to $\succ$.

   (3) A minimal Gröbner basis is **reduced** if no non-initial term of any g_i is divisible by any of $\mathrm{in}_\succ(g_1), \ldots, \mathrm{in}_\succ(g_s)$.

Example 12.4. Let $I = \langle x^2 - y, x^3 - x \rangle$. Then with respect to the lex order with $x \succ y$, the reduced Gröbner basis of I is

$$\{y^2 - y, xy - x, x^2 - y\}.$$

We will see later how to compute this. In particular, this implies that $\mathrm{in}_\succ(I) = \langle x^2, xy, y^2 \rangle$. The set $\{y^2 - y, xy - x + y^2 - y, x^2 - y\}$ is a minimal Gröbner basis of I with respect to the above lex order while $\{y^2 - y, xy - x + y^2 - y, x^2 - y, x^3 - x\}$ is a non-minimal Gröbner basis of I with respect to the same order.

As you can see in the above example, Gröbner bases of ideals are not unique. However, the reduced Gröbner basis of I with respect to a fixed term order $\succ$ is unique provided we assume that the polynomials in the reduced Gröbner basis are monic.

Lemma 12.5. *A Gröbner basis* $\mathcal{G}_\succ(I)$ *of* I *is a basis of* I.

Proof. By definition, $\langle \mathcal{G}_\succ(I) \rangle \subseteq I$. So we need to show that if $f \in I$, then $f \in \langle \mathcal{G}_\succ(I) \rangle$. Suppose not. Then we can assume without loss of generality that f is monic and that among all polynomials of I that are not in $\langle \mathcal{G}_\succ(I) \rangle$, f has the smallest initial monomial $\text{in}_\succ(f)$ with respect to $\succ$. (Such an f is called a *minimal criminal*.) However, $f \in I$ implies that $\text{in}_\succ(f) \in \text{in}_\succ(I)$, which implies that there exists some $g \in \mathcal{G}_\succ(I)$ such that $\text{in}_\succ(g)$ divides $\text{in}_\succ(f)$. Suppose $\text{in}_\succ(g) \cdot \mathbf{x^m} = \text{in}_\succ(f)$. Then $h = f - \mathbf{x^m} \cdot g$ is a polynomial in I with smaller initial term than $\text{in}_\succ(f)$. By our assumption, $h \in \langle \mathcal{G}_\succ(I) \rangle$, which implies that $f \in \langle \mathcal{G}_\succ(I) \rangle$, which is a contradiction. Thus $I \subseteq \langle \mathcal{G}_\succ(I) \rangle$. $\square$

Theorem 12.6. Hilbert's Basis Theorem. *Every ideal* I *in* S *has a finite generating set.*

Proof. By Lemma 12.5, a reduced Gröbner basis of I with respect to any term order $\succ$ is a basis of I. This reduced Gröbner basis is finite as it contains as many polynomials as the unique finite generating set of the monomial ideal $\text{in}_\succ(I)$. In a previous exercise, we proved that all monomial ideals have a unique finite generating set. $\square$

Lemma 12.7. *If* $\mathcal{G}_\succ(I)$ *is a Gröbner basis of* I *with respect to the term order* $\succ$, *then the remainder of any polynomial after division by* $\mathcal{G}_\succ(I)$ *is unique.*

Proof. Suppose $\mathcal{G}_\succ(I) = \{g_1, \ldots, g_s\}$ and suppose that we divide a polynomial $f \in S$ by $\mathcal{G}_\succ(I)$ with the elements of $\mathcal{G}_\succ(I)$ ordered in two different ways and that we obtain two remainders $r_1, r_2 \in S$. Then we have the two expressions

$$f = \sum a_i g_i + r_1 = \sum a_i' g_i + r_2$$

which implies that $r_1 - r_2 \in I$ and that no term of $r_1 - r_2$ is divisible by $\text{in}_\succ(g_i)$ for any $g_i \in \mathcal{G}_\succ(I)$. However this implies that $r_1 - r_2 = 0$ since otherwise the non-zero term $\text{in}_\succ(r_1 - r_2) \in \text{in}_\succ(I)$ and some $\text{in}_\succ(g_i)$ would divide it. $\square$

Corollary 12.8. *Gröbner bases solve the ideal membership problem: A polynomial* f *is in* I *if and only if its remainder after division by a Gröbner basis* $\mathcal{G}_\succ(I)$ *is zero.*

Proof. If $f \in S$ and its remainder after division by $\mathcal{G}_\succ(I)$ is zero, then $f = \sum_{g_i \in \mathcal{G}_\succ(I)} h_i g_i$ which lies in I, and hence, $f \in I$.

Conversely, suppose $f \in I$. Then since $\mathcal{G}_\succ(I)$ is a basis of I, there exists $h_i \in S$ such that $f = \sum_{g_i \in \mathcal{G}_\succ(I)} h_i g_i$. Then $0 = f - \sum_{g_i \in \mathcal{G}_\succ(I)} h_i g_i$ which implies that 0 is one, and thus the unique, remainder of f on division by $\mathcal{G}_\succ(I)$. □

In [**Buc65**], Buchberger developed an algorithm to compute a reduced Gröbner basis of an ideal $I = \langle f_1, \ldots, f_t \rangle$ with respect to any prescribed term order $\succ$ on S. The algorithm needs as a subroutine the calculation of the S-pair of two polynomials f and g, denoted as S-pair(f, g). Let $\mathbf{x}^\gamma$ be the least common multiple of $\mathrm{LT}_\succ(f)$ and $\mathrm{LT}_\succ(g)$. Then

$$\text{S-pair}(f, g) = \frac{\mathbf{x}^\gamma}{\mathrm{LT}_\succ(f)} \cdot f - \frac{\mathbf{x}^\gamma}{\mathrm{LT}_\succ(g)} \cdot g.$$

We also let $\mathrm{rem}_G(h)$ denote the remainder obtained by dividing the polynomial h by an ordered list of polynomials G.

Buchberger's algorithm hinges on the important fact that a set of polynomials G forms a Gröbner basis with respect to $\succ$ if and only if for each pair $f, f' \in G$, $\mathrm{rem}_G(\text{S-pair}(f, f')) = 0$. The proof can be found in any of the books mentioned at the start of Chapter 11. We reproduce the algorithm from [**CLO92**, §2.7, Theorem 2].

Algorithm 12.9. Buchberger's algorithm.
INPUT: $F = \{f_1, \ldots, f_t\}$, a basis of the ideal $I \subset S$, and a term order $\succ$ on S.
OUTPUT: The reduced Gröbner basis $\mathcal{G}_\succ(I)$ of I with respect to $\succ$.
$G := F$
REPEAT
    $G' := G$
    For each pair $\{p, q\}$, $p \neq q$ in G', do
        $S := \mathrm{rem}_{G'}(\text{S-pair}(p, q))$
        If $S \neq 0$, then $G := G \cup \{S\}$
UNTIL $G = G'$.
(G is a Gröbner basis of I with respect to $\succ$ at this point.)

Producing a minimal Gröbner basis.
Make every element of G monic by dividing through by its leading coefficient. Let $U := \{\mathrm{in}_{\succ}(g) : g \in G\}$. For each minimal element of U with respect to divisibility, pick one polynomial from G whose initial monomial is this minimal element. Call this set G again.

Producing the reduced Gröbner basis.
Let G be a minimal Gröbner basis of I with respect to $\succ$.
Set $G' := G$ and $\mathcal{G}_{\succ}(I) := \emptyset$.
For each $g \in G$ do
$$g' = \mathrm{rem}_{G' \setminus \{g\}}(g); \ \mathcal{G}_{\succ}(I) = \mathcal{G}_{\succ}(I) \cup \{g'\}; \ G' = G' \setminus \{g\} \cup \{g'\}.$$

Example 12.10. For the ideal $I = \langle f_1 := x^2 - y, f_2 := x^3 - x \rangle$ with the lex order $x \succ y$, we begin by setting $G = \{f_1, f_2\}$. The first step of the Buchberger algorithm computes S-pair$(f_1, f_2) = f_2 - x(f_1) = xy - x$. Note that $\mathrm{rem}_G(xy - x) = xy - x$. Thus we define $f_3 := xy - x$ and update G to $G = \{f_1, f_2, f_3\}$. Continue the Buchberger algorithm until all S-pairs reduce to zero, at which stage we will have $G = \{f_1, f_2, f_3\}$. The reduced Gröbner basis of I with respect to $\succ$ is thus $\{x^2 - y, xy - x\}$.

Definition 12.11. A finite set $\mathcal{U} \subset I$ is a **universal Gröbner basis** of I if it is a Gröbner basis of I with respect to *every* term order.

Proposition 12.12. Linear ideals revisited (cf. Chapter 11).

(1) *The set of linear forms $g_1, \ldots, g_d$ computed from the Gauss-Jordan form C of the matrix A is the reduced Gröbner basis of the linear ideal I of A with respect to any term order such that $x_1 \succ \cdots \succ x_n$.*

(2) *The set of all circuits of I is a minimal universal Gröbner basis of I ([**Stu96**, Proposition 1.6]).*

Proof.   (1) We follow the notation in Proposition 11.13. Note that the terms of each $g_i = x_i + \sum_{j=d+1}^{n} e_{ij} x_j$ are already ordered in decreasing order with respect to the above term order and that $G = \{g_1, \ldots, g_d\}$ is reduced in the sense that no term of any g_i other than x_i lies in the initial ideal $\mathrm{in}_{\succ}(I) = \langle x_1, \ldots, x_d \rangle$. To show that G is a Gröbner basis,

it suffices to show that the remainder obtained by dividing
S-pair(g_i, g_j) with respect to G is zero for all $i \neq j \in [d]$.
This follows from the following general fact: If $f = \mathrm{in}_\succ(f) + f'$ and $g = \mathrm{in}_\succ(g) + g'$ are two monic polynomials such that
$\mathrm{in}_\succ(f)$ and $\mathrm{in}_\succ(g)$ are relatively prime, then

$$\text{S-pair}(f, g) = \mathrm{in}_\succ(g) \cdot f - \mathrm{in}_\succ(f) \cdot g,$$

which reduces to zero modulo $\{f, g\}$. This is an important
criterion for avoiding S-pairs that will eventually reduce to
zero, known as *Buchberger's first criterion*.

(2) (Proof from [**Stu96**]) The argument in (1) shows that every
reduced Gröbner basis of I arises from a Gauss-Jordan form.
Proposition 11.13(2) proved that all the elements of these
Gröbner bases are circuits of I. Thus the circuits of I form
a universal Gröbner basis of I.

To prove minimality, we need to argue that each circuit
l appears in some reduced Gröbner basis of I. Let $\succ$ be a
term order such that $\{x_i : i \notin \mathrm{supp}(l)\} \succ \{x_i : i \in \mathrm{supp}(l)\}$
and $\mathcal{G} := \mathcal{G}_\succ(I)$. Such term orders are known as *elimination orders*. Suppose l does not appear in $\mathcal{G}$. Then
there exists $l' \in \mathcal{G}$ such that $\mathrm{in}_\succ(l) = \mathrm{in}_\succ(l')$. By the
elimination property of $\succ$, $\mathrm{supp}(l') \subseteq \mathrm{supp}(l)$ and hence
$\mathrm{supp}(l - l') \subsetneq \mathrm{supp}(l)$. However this contradicts the fact
that l is a circuit of I as $l - l'$ is a non-zero linear form with
strictly smaller support.

$\square$

Exercise 12.13. Fixing a term order, use Buchberger's algorithm to
confirm that $x^3 - y^2 \in \langle x - y, \, x^2 - y \rangle$.

Exercise 12.14. Let $M = \langle \mathbf{x}^{\mathbf{m}_1}, \mathbf{x}^{\mathbf{m}_2}, \ldots, \mathbf{x}^{\mathbf{m}_s} \rangle$ and fix a term order
$\succ$. What is a Gröbner basis for M with respect to $\succ$? What would
you expect its reduced Gröbner basis to be? Should your choice of
term order matter?

Chapter 13

Initial Complexes of Toric Ideals

In this chapter we will set the stage for another polytope, called the *state polytope* of a point configuration. The state polytope is *finer* than the secondary polytope in a precise sense. For this we first pass to a polynomial ideal called the *toric ideal* of the configuration and then to its Gröbner bases.

Let $\mathcal{A} \subset \mathbb{Z}^d$ be a graded vector configuration and A the $d \times n$ matrix whose columns are the elements of $\mathcal{A}$. We will assume as usual that $\text{rank}(A) = d$ and that $(1, 1, \ldots, 1)$ lies in the row space of A. Let

$$\ker_{\mathbb{Z}}(A) = \{\mathbf{u} \in \mathbb{Z}^n : A\mathbf{u} = \mathbf{0}\}.$$

Notice that $\ker_{\mathbb{Z}}(A)$ is an abelian subgroup of $\mathbb{Z}^n$. We can write a vector $\mathbf{u} \in \mathbb{Z}^n$ uniquely as $\mathbf{u} = \mathbf{u}^+ - \mathbf{u}^-$ where $\mathbf{u}^+, \mathbf{u}^- \in \mathbb{N}^n$ and

$$(\mathbf{u}^+)_j = \begin{cases} u_j & \text{if } u_j \geq 0 \\ 0 & \text{otherwise,} \end{cases}$$

$$(\mathbf{u}^-)_j = \begin{cases} -u_j & \text{if } u_j \leq 0 \\ 0 & \text{otherwise.} \end{cases}$$

Example 13.1. If $\mathbf{u} = (-5, 16, 0, 0, -17, 2)$, then $\mathbf{u}^+ = (0, 16, 0, 0, 0, 2)$ and $\mathbf{u}^- = (5, 0, 0, 0, 17, 0)$.

Definition 13.2. The **toric ideal** of $\mathcal{A}$ is the polynomial ideal

$$I_{\mathcal{A}} := \langle \mathbf{x}^{\mathbf{u}^+} - \mathbf{x}^{\mathbf{u}^-} \; : \; \mathbf{u} \in \ker_{\mathbb{Z}}(A) \rangle \subset \mathbf{k}[x_1, \ldots, x_n] =: S.$$

Definition 13.2 defines $I_{\mathcal{A}}$ via an infinite generating set consisting of **binomials**. Binomials are polynomials with two terms, just as monomials are polynomials with one term. However, by Hilbert's basis theorem, we know that there is a finite subset of the above infinite generating set that also generates $I_{\mathcal{A}}$. There are algorithms for finding such a finite generating set, although it becomes a highly nontrivial calculation as the size of $\mathcal{A}$ increases. See [**Stu96**, Chapters 4 and 12.A] if you are interested in these algorithms. The following Macaulay 2 session (from [**SST02**]) implements one of these algorithms and computes the toric ideal of the configuration $\mathcal{A}$ consisting of the columns of

$$\begin{pmatrix} 4 & 0 & 0 & 2 & 1 & 1 \\ 0 & 4 & 0 & 1 & 2 & 1 \\ 0 & 0 & 4 & 1 & 1 & 2 \end{pmatrix}.$$

We will use it as a black box to generate toric ideals.

Example 13.3. `[thomas@penguin thomas]$ M2`

```
Macaulay 2, version 0.9.2
i1 : load "LLL.m2"

i2 : toBinomial = (b,R) -> (
   top := 1_R; bottom := 1_R;
   scan(#b, i -> if b_i > 0 then top = top * R_i^(b_i)
   else if b_i < 0 then bottom = bottom * R_i^(-b_i));
   top - bottom);

i3 : toricIdeal = (A) -> (
   n := #(A_0);
   R = QQ[vars(0..n-1),MonomialSize=>16];
   B := transpose matrix LLL syz matrix A;
   J := ideal apply(entries B, b -> toBinomial(b,R));
   scan(gens ring J, f -> (J = saturate(J,f);));
   J
   );
```

```
i4 : A = {{4,0,0,2,1,1},{0,4,0,1,2,1},{0,0,4,1,1,2}};

i5 : I = toricIdeal A
             3                    3   3
o5 = ideal (e  - b*d*f, c*d*e - f , d  - a*e*f, ...

o5 : Ideal of R
i6 : toString oo
o6 = ideal(e^3-b*d*f,c*d*e-f^3,d^3-a*e*f,a*b*c-d*e*f,
d^2*e^2-a*b*f^2,a*c*e^2-d^2*f^2,b*c*d^2-e^2*f^2)
```

The output on line o6 is the finite generating set of I_A.

A **grading** of S by an abelian group G is a map called *degree* defined as deg : $S \to G$ such that $x_i \mapsto g_i$ for each $i = 1, \ldots, n$. This map is extended to a monomial $\mathbf{x^a} \in S$ as $\deg(\mathbf{x^a}) = \sum a_i g_i$. A polynomial $f \in S$ is **homogeneous** under this grading if all monomials in f have the same degree. An ideal $I \subset S$ is homogeneous if every polynomial in I is a sum of homogeneous polynomials or, equivalently, if I is generated by homogeneous polynomials under the given grading. The most common grading on S is by $\mathbb{Z}$ where $\deg(x_i) = 1$. This is called the **total degree grading** and unless stated otherwise, we will mean this grading when we talk about homogeneous ideals.

Every homogeneous polynomial ideal has a state polytope which is a polytope whose vertices are in bijection with the distinct reduced Gröbner bases (equivalently, initial ideals) of the ideal. In the next chapter we will examine the state polytopes of toric ideals and see how they are related to secondary polytopes. For such a relationship to exist, we need a relationship between the initial ideals of the toric ideal of $\mathcal{A}$ and the regular triangulations of $\mathcal{A}$. We establish this connection in the rest of this chapter.

Definition 13.4. A **simplicial complex** Γ on $[n]$ is a collection of subsets of $[n]$ such that if $F \in \Gamma$ and $G \subset F$, then $G \in \Gamma$ as well. In particular, the empty set is in Γ.

    (1) The elements of Γ are called the **faces** of Γ.

(2) The dimension of a face $F \in \Gamma$ is $|F| - 1$. The dimension of Γ is the maximum dimension of a face in Γ.

(3) The maximal-dimensional faces of Γ are called facets.

(4) Γ is **pure** if all its facets have the same dimension.

(5) The face poset of Γ is the set of faces of Γ partially ordered by inclusion.

(6) A **non-face** of Γ is any set $H \subseteq [n]$ such that H is not a face of Γ. A **minimal non-face** of Γ is a non-face H of Γ such that all its proper subsets are faces of Γ.

Example 13.5. A triangulation of $\mathcal{A}$ gives a simplicial complex on $[n]$. For instance our favorite non-regular triangulation in Figure 1 gives the pure simplicial complex on $[6] = \{1, 2, 3, 4, 5, 6\}$ whose face lattice can also be seen in Figure 1.

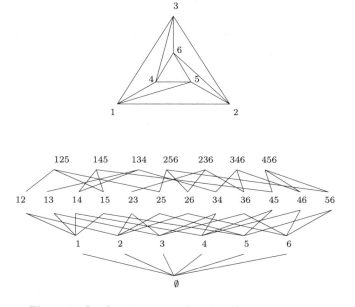

Figure 1. Our favorite non-regular triangulation as a simplicial complex.

The set $\{1, 2, 6\}$ is a non-minimal non-face of this simplicial complex while $\{1, 2, 3\}$ is a minimal non-face.

Note that a subdivision of $\mathcal{A}$ that is not a triangulation does not yield a simplicial complex, as only some subsets of a face of the subdivision may be faces of the subdivision. We will identify a triangulation with its associated simplicial complex. Note also that it is enough to specify the facets of a simplicial complex as the remaining faces are simply all subsets of the facets. Thus from now on we specify simplicial complexes by listing only their facets.

Given an ideal $I \subset S$ and a term order $\succ$, we can compute the initial ideal $\mathrm{in}_\succ(I)$ which is a monomial ideal in S. We saw earlier that the radical $\sqrt{\mathrm{in}_\succ(I)}$ is also a monomial ideal in S and is generated by squarefree monomials obtained by erasing the powers on the monomial generators of $\mathrm{in}_\succ(I)$. There is a bijection between squarefree monomial ideals in S and simplicial complexes on $[n]$.

Definition 13.6. The **Stanley-Reisner ideal** of a simplicial complex Γ on $[n]$ is the squarefree monomial ideal

$$I_\Gamma := \langle x_{i_1} x_{i_2} \cdots x_{i_k} : \{i_1, i_2, \ldots, i_k\} \text{ is a minimal non-face of } \Gamma \rangle.$$

Conversely, a squarefree monomial ideal I is the Stanley-Reisner ideal of the unique simplicial complex $\Gamma(I)$ whose set of minimal non-faces is

$$\{\{i_1, i_2, \ldots, i_k\} \subseteq [n] : x_{i_1} x_{i_2} \cdots x_{i_k} \text{ is a minimal generator of } I\}.$$

Example 13.7. The minimal non-faces of the non-regular triangulation in Figure 1 are $16, 35, 24, 123$ which implies that its Stanley-Reisner ideal is $\langle af, ce, bd, abc \rangle$ where we take $S = \mathbf{k}[a, b, c, d, e, f]$.

Exercise 13.8. (1) Prove that if $\{i_1, \ldots, i_k\}$ is any non-face of Γ, then $x_{i_1} x_{i_2} \cdots x_{i_k} \in I_\Gamma$.

(2) Finding all minimal non-faces of a simplicial complex is not so easy. It is easier to compute I_Γ via the following formula:

$$I_\Gamma = \bigcap_{\sigma \text{ facet of } \Gamma} \langle x_i : i \notin \sigma \rangle.$$

Check this formula for the Stanley-Reisner ideal constructed in Example 13.7 and then prove it in general.

If we have a non-squarefree monomial ideal $I \subset S$, then the simplicial complex we want to associate with it is $\Gamma(\sqrt{I})$. However, since many ideals can have the same radical, the map from monomial ideals to simplicial complexes is not a bijection. We will study simplicial complexes that come from initial ideals of a polynomial ideal.

Definition 13.9. The **initial complex** $\Gamma_\succ(I)$ of a polynomial ideal $I \subseteq S$ with respect to the term order $\succ$ is the simplicial complex $\Gamma(\sqrt{\text{in}_\succ(I)})$.

Example 13.10. Consider the vector configuration

$$\mathcal{A} = \left\{ \begin{pmatrix} 1 \\ 0 \end{pmatrix}, \begin{pmatrix} 1 \\ 1 \end{pmatrix}, \begin{pmatrix} 1 \\ 2 \end{pmatrix}, \begin{pmatrix} 1 \\ 3 \end{pmatrix} \right\}$$

from Example 7.11. Its toric ideal is

$$I_\mathcal{A} = \langle ac - b^2, bd - c^2, ad - bc \rangle \subset \mathbf{k}[a, b, c, d].$$

This ideal has eight distinct monomial initial ideals that are listed below along with weight vectors $\omega \in \mathbb{R}^4$ that induce them, their radicals, and initial complexes. On the extreme right we also list the regular triangulations Δ_ω of the configuration.

ω	$\text{in}_\omega(I)$	$\sqrt{\text{in}_\omega(I)}$	$\Gamma_\omega(I)$	Δ_ω
$(0,1,1,0)$	$\langle c^2, bc, b^2 \rangle$	$\langle b, c \rangle$	$\{14\}$	$\{14\}$
$(0,5,1,0)$	$\langle b^2, bc, c^3, bd \rangle$	"	"	"
$(0,1,5,0)$	$\langle b^3, ac, bc, c^2 \rangle$	"	"	"
$(0,3,0,1)$	$\langle b^2, bc, bd, ad^2 \rangle$	$\langle b, ad \rangle$	$\{13, 34\}$	$\{13, 34\}$
$(0,1,0,3)$	$\langle b^2, ad, bd \rangle$	"	"	"
$(1,0,0,1)$	$\langle ac, ad, bd \rangle$	$\langle ac, ad, bd \rangle$	$\{12, 23, 34\}$	$\{12, 23, 34\}$
$(1,0,3,0)$	$\langle ac, bc, c^2, a^2d \rangle$	$\langle c, ad \rangle$	$\{12, 24\}$	$\{12, 24\}$
$(3,0,1,0)$	$\langle ac, c^2, ad \rangle$	"	"	"

From this example, you might suspect that the initial complex $\Gamma_\omega(I_\mathcal{A})$ is precisely the regular triangulation Δ_ω. This is the main result of this section, which we prove in Theorem 13.12.

Lemma 13.11. *Let Δ_ω be a regular triangulation of $\mathcal{A}$, let $\sigma = \{i_1, \dots, i_k\}$ be a face of Δ_ω and let $\tau = \{j_1, \dots, j_l\}$ be a non-face of Δ_ω such that the relative interiors of their convex hulls intersect. Pick positive rational numbers $\lambda_1, \dots, \lambda_k$ and $\mu_1, \dots, \mu_l$ such that*

$$\lambda_1 \mathbf{a}_{i_1} + \dots + \lambda_k \mathbf{a}_{i_k} = \mu_1 \mathbf{a}_{j_1} + \dots + \mu_l \mathbf{a}_{j_l}$$

and $\sum \lambda_i = \sum \mu_i = 1$. Then $\sum_{i=1}^n \omega_i \lambda_i < \sum_{i=1}^n \omega_i \mu_i$.

Proof. Recall that the facets of Δ_w are precisely the lower facets of the lifted polytope P^ω. Therefore, σ lifts to a lower face of P^ω where for each $j \in \sigma$, $\mathbf{a}_j$ has been lifted to height ω_j. This implies that the convex combination $\lambda_1 \mathbf{a}_{i_1} + \ldots + \lambda_k \mathbf{a}_{i_k}$ has been lifted to height $\sum_{i \in \sigma} \lambda_i \omega_i$ since the lifted points indexed by σ span an affine plane. On the other hand, the convex hull of the points indexed by τ do not form a lower face of Δ_ω and hence, $\sum_{i \in \tau} \omega_i \mu_i > \sum_{i \in \sigma} \omega_i \lambda_i$. $\qquad\square$

Theorem 13.12. ([**Stu91**], Theorem 3.1]) *Let $I_\mathcal{A}$ be the toric ideal of $\mathcal{A}$ and let ω be a term order on S. Then the initial complex $\Gamma_\omega(I_\mathcal{A})$ is the regular triangulation Δ_ω of $\mathcal{A}$ or, equivalently, $\sqrt{\mathrm{in}_\omega(I_\mathcal{A})}$ is the Stanley-Reisner ideal I_{Δ_ω} of the regular triangulation Δ_ω of $\mathcal{A}$.*

Proof. We reproduce the proof in [**Stu91**]. Suppose Δ_ω is a regular triangulation of $\mathcal{A}$. We will first argue that the Stanley-Reisner ideal I_{Δ_ω} is contained in $\sqrt{\mathrm{in}_\omega(I_\mathcal{A})}$. Let $\tau = \{j_1, \ldots, j_l\}$ be a non-face of Δ_ω or, equivalently, $x_{j_1} x_{j_2} \cdots x_{j_l} \in I_{\Delta_\omega}$. Then there exists a face $\sigma = \{i_1, \ldots, i_k\}$ of Δ_ω such that

$$\mathrm{relint}(\mathrm{conv}(\mathbf{a}_{i_1}, \ldots, \mathbf{a}_{i_k})) \cap \mathrm{relint}(\mathrm{conv}(\mathbf{a}_{j_1}, \ldots, \mathbf{a}_{j_l})) \neq \emptyset.$$

Pick positive rational numbers $\lambda_1', \ldots, \lambda_k'$ and $\mu_1', \ldots, \mu_l'$ with $\sum \lambda_i' = \sum \mu_i' = 1$ such that

$$\lambda_1' \mathbf{a}_{i_1} + \ldots + \lambda_k' \mathbf{a}_{i_k} = \mu_1' \mathbf{a}_{j_1} + \ldots + \mu_l' \mathbf{a}_{j_l}.$$

Clearing denominators, we get vectors $\lambda, \mu \in \mathbb{N}^n$ such that

$$A\lambda = \lambda_{i_1} \mathbf{a}_{i_1} + \ldots + \lambda_{i_k} \mathbf{a}_{i_k} = \mu_{j_1} \mathbf{a}_{j_1} + \ldots + \mu_{j_l} \mathbf{a}_{j_l} = A\mu.$$

Then the vector $\pm(\lambda - \mu) \in \ker_\mathbb{Z}(A)$ which means that $\pm(\mathbf{x}^\lambda - \mathbf{x}^\mu)$ lies in the toric ideal $I_\mathcal{A}$. Since σ is a face of Δ_ω while τ is a non-face, by Lemma 13.11, we have that $\omega \cdot \lambda < \omega \cdot \mu$ which implies that $\mathrm{in}_\omega(\mathbf{x}^\lambda - \mathbf{x}^\mu) = \mathbf{x}^\mu$. This means that $\mathbf{x}^\mu \in \mathrm{in}_\omega(I_\mathcal{A})$ and subsequently, $x_{j_1} \cdots x_{j_l} \in \sqrt{\mathrm{in}_\omega(I_\mathcal{A})}$. This proves that $I_{\Delta_\omega} \subseteq \sqrt{\mathrm{in}_\omega(I_\mathcal{A})}$.

Conversely, suppose $\sqrt{\mathrm{in}_\omega(I_\mathcal{A})}$ is not contained in I_{Δ_ω}. This implies that there is a monomial $\mathbf{x}^\mu$ in $\mathrm{in}_\omega(I_\mathcal{A})$ whose support is a face of Δ_ω. It is not too hard to prove that the ideal $I_\mathcal{A}$ is generated as a $\mathbf{k}$-vector space by binomials of the form $\mathbf{x}^{\mathbf{u}^+} - \mathbf{x}^{\mathbf{u}^-}$ where $\mathbf{u} \in \ker_\mathbb{Z}(A)$ which then means that there is a binomial $\mathbf{x}^\mu - \mathbf{x}^\lambda \in I_\mathcal{A}$ with $\mathrm{in}_\omega(\mathbf{x}^\mu - \mathbf{x}^\lambda) = \mathbf{x}^\mu$ such that $\mathrm{supp}(\mathbf{x}^\mu)$ is a face of Δ_ω. We have

$\omega \cdot \mu > \omega \cdot \lambda$. Also, $A\lambda = A\mu$ which means that supp($\mathbf{x}^\lambda$) has to be a non-face of Δ_ω since otherwise we have two faces intersecting in their relative interiors. This contradicts Lemma 13.11, and hence, $\sqrt{\text{in}_\omega(I_\mathcal{A})} \subseteq I_{\Delta_\omega}$. □

Corollary 13.13. *A monomial* $\mathbf{x}^\mathbf{m}$ *is standard for the initial ideal* $\text{in}_\omega(I_\mathcal{A})$ *if and only if* supp($\mathbf{x}^\mathbf{m}$) *does not contain a non-face of* Δ_ω.

The toric ideal of the six-point configuration underlying our favorite non-regular triangulation has 112 initial ideals, which is a bit too large for us to analyze by hand. Instead, let's look at all the initial ideals of the pentagonal configuration from Example 8.3.

Exercise 13.14. Consider the pentagonal configuration $\mathcal{A}$ whose elements are the columns of

$$\begin{pmatrix} 1 & 1 & 1 & 1 & 1 \\ 0 & 1 & 2 & 1 & 0 \\ 0 & 0 & 1 & 2 & 1 \end{pmatrix}.$$

The toric ideal $I_\mathcal{A} = \langle bd - ce, a^2d - be^2, a^2c - b^2e \rangle \subset \mathbf{k}[a, b, c, d, e]$. Listed below are the eight distinct monomial initial ideals of $I_\mathcal{A}$:

(1) $\langle bd, a^2d, a^2c \rangle$,

(2) $\langle ce, a^2d, a^2c \rangle$,

(3) $\langle ce, be^2, a^2c \rangle$,

(4) $\langle ce, be^2, b^2e, a^2c^2 \rangle$,

(5) $\langle bd, b^2e, a^2d \rangle$,

(6) $\langle be^2, bd, b^2e, a^2d^2 \rangle$,

(7) $\langle ce^3, be^2, bd, b^2e \rangle$,

(8) $\langle ce, be^2, b^2e, b^3d \rangle$.

This calculation was done using the software package CaTS [**Jena**]. Group these initial ideals by their radicals and compute the corresponding initial complexes. Check that they coincide with the five regular triangulations of this configuration that we saw earlier. Can you find weight vectors that induce each of the initial ideals?

Theorem 13.12 says that regular triangulations support initial ideals in the sense that their Stanley-Reisner ideals are radicals of initial ideals of I_A. Equivalently, the set of regular triangulations of $\mathcal{A}$ is precisely the set of simplicial complexes coming from radicals of initial ideals of I_A. Thus a non-regular triangulation cannot support an initial ideal of I_A. Let's prove this for our favorite regular triangulation.

Theorem 13.15. *There is no weight vector ω such that the initial complex of I_A with respect to ω is our favorite non-regular triangulation.*

Proof. Look at the three quadrilaterals $\{1, 2, 4, 5\}, \{1, 3, 4, 6\}$ and $\{2, 3, 5, 6\}$ in the non-regular triangulation. These quadrilaterals support the dependence relations

$$(-1, 1, 0, 4, -4, 0), (1, 0, -1, -4, 0, 4) \text{ and } (0, -1, 1, 0, 4, -4)$$

on the columns of

$$A = \begin{pmatrix} 4 & 0 & 0 & 2 & 1 & 1 \\ 0 & 4 & 0 & 1 & 2 & 1 \\ 0 & 0 & 4 & 1 & 1 & 2 \end{pmatrix}.$$

These elements in $\ker_{\mathbb{Z}}(A)$ give the following three binomials in the toric ideal I_A:

$$bd^4 - ae^4, af^4 - cd^4, ce^4 - bf^4.$$

In each case, the positive monomial is the leading term with respect to any ω that induces this triangulation. This is because the support of the positive term is a non-face of the triangulation while the support of the other monomial is a face of the triangulation. This implies the following three inequalities:

$$\omega_2 + 4\omega_4 > \omega_1 + 4\omega_5,$$
$$\omega_1 + 4\omega_6 > \omega_3 + 4\omega_4,$$
$$\omega_3 + 4\omega_5 > \omega_2 + 4\omega_6.$$

However, if you add the three inequalities, you get $0 > 0$, which implies that no such ω exists. □

Chapter 14

State Polytopes of Toric Ideals

In this chapter we will define and construct the *state polytope* of a toric ideal $I_{\mathcal{A}}$ and then relate it to the secondary polytope of $\mathcal{A}$ via the relationship between initial ideals of $I_{\mathcal{A}}$ and regular triangulations of $\mathcal{A}$ developed in Chapter 13. We follow the same approach as with secondary polytopes — we will first define a polyhedral fan called the *Gröbner fan* of $I_{\mathcal{A}}$ which will then be shown to be polytopal. This polytope will be called the state polytope of $I_{\mathcal{A}}$. State polytopes and Gröbner fans exist for all homogeneous ideals [**BM98, MR88**] and [**Stu96**, Chapter 2]. For non-homogeneous ideals, one can define the Gröbner fan, but these fans may not be polytopal. An example of such an ideal can be found in [**Jenb**].

Before we discuss the general situation, let us take a homogeneous principal ideal and work out its state polytope. The vertices of this polytope must index the distinct reduced Gröbner bases of the ideal. If $I = \langle f \rangle$ where f is homogeneous, then all reduced Gröbner bases of I consist of the single element f with different terms marked as leading monomial. Which monomials in the support of f can become leading terms? For a given weight vector ω, these are precisely the monomials $\mathbf{x}^{\alpha}$ in the support of f for which $\alpha \cdot \omega$ is maximized. Thus the exponents of the possible initial monomials of f are in bijection with the vertices of the **Newton polytope** of f which is the convex hull of all the exponent vectors of monomials in f. The state polytope of $I = \langle f \rangle$ is the Newton polytope of f.

Example 14.1. Let $I = \langle f = 3x^3 + \frac{1}{2}x^2y + \sqrt{2}xy^2 + y^3 \rangle$. Then the Newton polytope of f is the line segment

$$\text{conv}(\{(3,0),(2,1),(1,2),(0,3)\}),$$

and the two vertices of this line segment correspond to the two distinct reduced Gröbner bases of I which are $\mathcal{G}_1 = \{\underline{x}^3 + \frac{1}{6}x^2y + \frac{\sqrt{2}}{3}xy^2 + \frac{1}{3}y^3\}$ and $\mathcal{G}_2 = \{3x^3 + \frac{1}{2}x^2y + \sqrt{2}xy^2 + \underline{y}^3\}$. The underlined term is the leading term in each case.

We will only be concerned with toric ideals. We assume the same setup as always: $\mathcal{A}$ is a graded point configuration in $\mathbb{Z}^d$ whose columns form the $d \times n$ integer matrix A of rank d. We assume that the vector $(1,1,\ldots,1)$ lies in the row space of A. Then the toric ideal of $\mathcal{A}$ is the homogeneous binomial ideal

$$I_{\mathcal{A}} = \langle \mathbf{x}^{\mathbf{u}^+} - \mathbf{x}^{\mathbf{u}^-} : \mathbf{u} \in \ker_{\mathbb{Z}}(A) \rangle.$$

Check that $I_{\mathcal{A}}$ is homogeneous.

Exercise 14.2. Argue that every reduced Gröbner basis $\mathcal{G}_\omega$ of the toric ideal $I_{\mathcal{A}}$ consists of a finite set of binomials of the form $\mathbf{x}^\alpha - \mathbf{x}^\beta$ such that $\alpha - \beta \in \ker_{\mathbb{Z}}(A)$.

(**Hint:** We showed earlier that $I_{\mathcal{A}}$ has a finite generating set of this form. If we use this as the input to Buchberger's algorithm, what sorts of polynomials will we generate during and at the end of the algorithm?)

Example 14.3. The eight reduced Gröbner bases of $I_{\mathcal{A}}$ for

$$\mathcal{A} = \left\{ \begin{pmatrix} 1 \\ 0 \end{pmatrix}, \begin{pmatrix} 1 \\ 1 \end{pmatrix}, \begin{pmatrix} 1 \\ 2 \end{pmatrix}, \begin{pmatrix} 1 \\ 3 \end{pmatrix} \right\}$$

are listed below. We saw these eight initial ideals in Example 13.10. They are

    (1) $\{b^2 - ac, bc - ad, c^2 - bd\}$,

    (2) $\{b^2 - ac, bc - ad, bd - c^2, c^3 - ad^2\}$,

    (3) $\{ac - b^2, b^3 - a^2d, bc - ad, c^2 - bd\}$,

(4) $\{ad^2 - c^3, b^2 - ac, bc - ad, bd - c^2\}$,

(5) $\{ad - bc, b^2 - ac, bd - c^2\}$,

(6) $\{ac - b^2, ad - bc, bd - c^2\}$,

(7) $\{a^2d - b^3, ac - b^2, bc - ad, c^2 - bd\}$,

(8) $\{ac - b^2, ad - bc, c^2 - bd\}$.

Definition 14.4. Let $\mathcal{G}_\omega = \{\mathbf{x}^{\alpha_i} - \mathbf{x}^{\beta_i} : i = 1, \ldots, t\}$ be a reduced Gröbner basis of I_A. The **Gröbner cone** of $\mathcal{G}_\omega$ is the polyhedral cone

$$\mathcal{K}_\omega = \{\mathbf{v} \in \mathbb{R}^n : \alpha_i \cdot \mathbf{v} \geq \beta_i \cdot \mathbf{v} \text{ for all } i = 1, \ldots, t\}.$$

The Gröbner cone $\mathcal{K}_\omega$ is full-dimensional (n-dimensional) since the weight vector ω satisfies the t inequalities with a strict inequality. Furthermore, the entire row space of A lies in $\mathcal{K}_\omega$ since $(\alpha_i - \beta_i) \in \ker_\mathbb{Z}(A)$ which implies that every vector in the row space of A will satisfy the t inequalities with equality. Is there a larger vector space in $\mathcal{K}_\omega$? If there is, then for every $\mathbf{v}$ in this vector space, $-\mathbf{v}$ is also in the vector space and we must have $\alpha_i \cdot \mathbf{v} \geq \beta_i \cdot \mathbf{v}$ and $\alpha_i \cdot -\mathbf{v} \geq \beta_i \cdot -\mathbf{v}$ which together imply that $(\alpha_i - \beta_i) \cdot \mathbf{v} = \mathbf{0}$. Now it turns out that the span of $\{\alpha_i - \beta_i : i = 1, \ldots, t\}$ is exactly $\ker_\mathbb{R}(A)$ which implies that the row space of A is the biggest vector space in $\mathcal{K}_\omega$. Thus we can write $\mathcal{K}_\omega$ as the sum of the row space of A and the part that lies in the kernel of A. The latter is the interesting part — it is a *pointed* polyhedral cone of dimension $n - d$, which we will call the **pointed Gröbner cone** of $\mathcal{G}_\omega$, denoted as $\mathcal{K}'_\omega$. We will now figure out how to draw $\mathcal{K}'_\omega$ in $\ker_\mathbb{R}(A)$ which is a vector space isomorphic to $\mathbb{R}^{n-d}$. From $\mathcal{K}'_\omega$ we can reconstruct $\mathcal{K}_\omega$ by adding on the row space of A.

Let $\mathcal{B}$ be the Gale transform of $\mathcal{A}$. Since $(1, 1, \ldots, 1)$ is already in the row space of A, the Gale transform $\mathcal{B} \subset \mathbb{R}^{n-d}$. Let B^t be the matrix whose columns give $\mathcal{B}$. Since A is an integer matrix, we can choose $B^t \in \mathbb{Z}^{(n-d) \times n}$ such that its rows generate $\ker_\mathbb{Z}(A)$ which in particular means that they also generate $\ker_\mathbb{R}(A)$. Furthermore, the row space of B^t is isomorphic to the column space of B^t which implies that the span of $\mathcal{B}$ is isomorphic to $\ker_\mathbb{R}(A)$ which in turn is isomorphic to $\mathbb{R}^{n-d}$. Thus we can draw $\mathcal{K}'_\omega$ in $\mathbb{R}^{n-d}$, the space spanned by the Gale transform $\mathcal{B}$. Given $\alpha - \beta \in \ker_\mathbb{R}(A)$, solve

$\mathbf{y}B^t = \alpha - \beta$ to find its representative $\mathbf{y} \in \mathbb{R}^{n-d}$. Let's use this method to draw the eight Gröbner cones of Example 14.3 in $\mathbb{R}^2$.

Example 14.5. For this example, we choose

$$B^t = \begin{pmatrix} 1 & -2 & 1 & 0 \\ 2 & -3 & 0 & 1 \end{pmatrix}.$$

Let $\mathcal{G}_\omega = \{\underline{b}^2 - ac, \underline{b}c - ad, \underline{c}^2 - bd\}$. Its Gröbner cone is

$$\mathcal{K}_\omega = \{\mathbf{v} \in \mathbb{R}^4 : 2v_2 \geq v_1 + v_3, \; v_2 + v_3 \geq v_1 + v_4, \; 2v_3 \geq v_2 + v_4\}.$$

We find the pre-images of the vectors $(-1, 2, -1, 0)$, $(-1, 1, 1, -1)$ and $(0, -1, 2, -1)$ under the map from $\mathbb{R}^2 \to \mathbb{R}^4$ where $\mathbf{y} \mapsto \mathbf{y}B^t$. Under this map, $(-1, 2, -1, 0) \mapsto (-1, 0)$, $(-1, 1, 1, -1) \mapsto (1, -1)$ and $(0, -1, 2, -1) \mapsto (2, -1)$. This implies that

$$\mathcal{K}'_\omega = \{(y_1, y_2) \in \mathbb{R}^2 : -y_1 \geq 0, y_1 - y_2 \geq 0, 2y_1 - y_2 \geq 0\}.$$

This cone is the shaded cone shown in Figure 1.

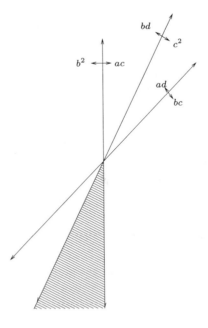

Figure 1. Pointed Gröbner cone.

We label the hyperplanes that bound the halfspaces with the monomials of the original binomial that gave rise to this hyperplane so that the monomial pointing into a halfspace is more expensive for a weight vector lifted from that region than the other monomial.

Using this technique, let us draw all eight Gröbner cones. The resulting picture is shown in Figure 3. To make our life easier, we first note that there are only five different binomials in the eight reduced Gröbner bases, up to sign. Thus we draw the hyperplanes corresponding to them first and mark the halfspaces with monomials as before. Then it is easy to pick off the Gröbner cones. Figure 2 shows the five hyperplanes with labeled halfspaces.

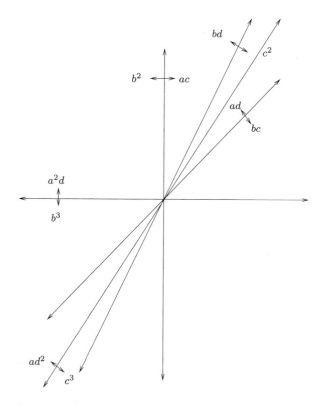

Figure 2. Hyperplanes given by binomials in the Gröbner bases.

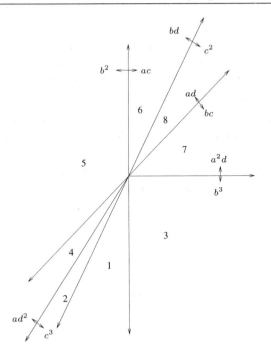

Figure 3. Pointed Gröbner fan.

Notice that the Gröbner cones fit together to form a polyhedral fan in the above example (see Figure 3). We will prove that this is always the case. In the meantime, let's denote by $\mathcal{GF}(\mathcal{A})$ the collection of all Gröbner cones of $I_\mathcal{A}$. We will call it the **Gröbner fan** of $I_\mathcal{A}$ although we have not yet proved that it is a fan.

Lemma 14.6. *Let ω be a term order for $I_\mathcal{A}$ and let $\mathcal{G}_\omega$ be the reduced Gröbner basis of $I_\mathcal{A}$ with respect to ω. Then $\mathcal{G}_{\omega'} = \mathcal{G}_\omega$ if and only if ω' lies in the interior of the Gröbner cone $\mathcal{K}_\omega$.*

Proof. Suppose $\mathcal{G}_\omega = \{\mathbf{x}^{\alpha_i} - \mathbf{x}^{\beta_i} : i = 1, \ldots, t\} = \mathcal{G}_{\omega'}$. Then $\alpha_i \cdot \omega' > \beta_i \cdot \omega'$ for all $i = 1, \ldots, t$ which implies that ω' lies in the interior of $\mathcal{K}_\omega$.

Conversely, suppose ω' lies in the interior of $\mathcal{K}_\omega$. Then $\alpha_i \cdot \omega' > \beta_i \cdot \omega'$ for all binomials $\mathbf{x}^{\alpha_i} - \mathbf{x}^{\beta_i}$ in $\mathcal{G}_\omega$. This implies that the initial ideal $\mathrm{in}_\omega(I_\mathcal{A}) \subseteq \mathrm{in}_{\omega'}(I_\mathcal{A})$. Therefore, the two initial ideals are equal since the standard monomials of both initial ideals form a **k**-vector space basis for S/I_A and if one initial ideal is properly inside another, then we would have one basis of this vector space properly inside another, which is a contradiction. Since $\mathcal{G}_\omega$ and $\mathcal{G}_{\omega'}$ are reduced, they are equal as every initial ideal gives rise to a unique reduced Gröbner basis of I_A. □

Corollary 14.7. *Two distinct Gröbner cones of I_A do not intersect in their interiors.*

We now construct a polytope whose inner normal fan is $\mathcal{GF}(A)$, proving that $\mathcal{GF}(A)$ is a polyhedral fan and that it is polytopal.

Definition 14.8. Take *the* universal Gröbner basis of I_A, denoted as $UGB(I_A)$, to be the set of all binomials (up to sign) that appear in the reduced Gröbner bases of I_A.

Example 14.9. In our running example, $UGB(I_A) = \{b^2 - ac, c^2 - bd, ad - bc, a^2d - b^3, ad^2 - c^3\}$.

It is a fact that every polynomial ideal has a finite universal Gröbner basis. This follows from the more important fact that every polynomial ideal has only finitely many distinct reduced Gröbner bases [**Stu96**, Chapter 1] and that each reduced Gröbner basis is finite. The union of these reduced Gröbner bases is always a universal Gröbner basis of the ideal.

We now define a grading of $S = \mathbf{k}[x_1, \ldots, x_n]$ as follows. Define $\deg(x_i) = \mathbf{a}_i \in A$ for $i = 1, \ldots, n$. Then $\deg(\mathbf{x}^{\mathbf{u}}) = A\mathbf{u} \in \mathbb{Z}^d$. We sometimes abuse notation and also refer to $A\mathbf{u}$ as the degree of $\mathbf{u} \in \mathbb{N}^n$. Notice that the two monomials in a binomial $\mathbf{x}^{\mathbf{u}^+} - \mathbf{x}^{\mathbf{u}^-} \in I_A$ have the same degree under this A-grading. For $\mathbf{u} \in \mathbb{N}^n$, let

$$\deg^{-1}(A\mathbf{u}) := \{\mathbf{v} \in \mathbb{N}^n : \deg(\mathbf{u}) = \deg(\mathbf{v})\}$$

be called the **fiber** of $\mathbf{u}$ or $\mathbf{x}^{\mathbf{u}}$.

Definition 14.10. A vector $\mathbf{b} \in \mathbb{Z}^d$ is a **Gröbner degree** of I_A if there is some element $\mathbf{x}^\alpha - \mathbf{x}^\beta \in UGB(I_A)$ such that $\mathbf{b} = \deg(\alpha) = \deg(\beta)$. The fibers of Gröbner degrees are called **Gröbner fibers**.

Example 14.11. The Gröbner degrees of our running example are

$$\{(2,2)^t, (2,4)^t, (2,3)^t, (3,3)^t, (3,6)^t\}.$$

The Gröbner fibers are

(1) $\deg^{-1}((2,2)^t) = \{(1,0,1,0),(0,2,0,0)\}$,

(2) $\deg^{-1}((2,4)^t) = \{(0,1,0,1),(0,0,2,0)\}$,

(3) $\deg^{-1}((2,3)^t) = \{(1,0,0,1),(0,1,1,0)\}$,

(4) $\deg^{-1}((3,3)^t) = \{(2,0,0,1),(1,1,1,0),(0,3,0,0)\}$,

(5) $\deg^{-1}((3,6)^t) = \{(1,0,0,2),(0,1,1,1),(0,0,3,0)\}$.

We can compute all monomials of a given degree using the following Macaulay 2 commands:

```
Macaulay 2, version 0.9.2

i1 : A = {{1,1,1,1},{0,1,2,3}};
i2 : R = QQ[a..d, Degrees => transpose A];
i3 : basis({2,2},R)
o3 = | ac b2 |
              1       2
o2 : Matrix R  <--- R
```

Definition 14.12. The **state polytope** of $I_{\mathcal{A}}$ is the Minkowski sum

$$St(I_{\mathcal{A}}) := \sum_{\mathbf{b} \text{ Gröbner degree}} \text{conv}(\deg^{-1}(\mathbf{b})).$$

Example 14.13. In our example, since the left-most 2×2 submatrix of A is non-singular, the projection of any $\mathbf{v} \in \mathbb{R}^4$ of a given degree to $(v_3, v_4) \in \mathbb{R}^2$ can be uniquely lifted back to $\mathbf{v}$. Thus, we can draw the projections of all Gröbner fibers into the x_3, x_4-plane and then compute the state polytope. The summands and the Minkowski sum are shown in Figure 4.

In Figure 5 we see that the inner normal fan of the state polytope coincides with the Gröbner fan we saw in Figure 3.

Theorem 14.14. *The Gröbner fan $\mathcal{GF}(\mathcal{A})$ of $I_{\mathcal{A}}$ is the inner normal fan of $St(I_{\mathcal{A}})$.*

Proof. Let ω be a term order for $I_{\mathcal{A}}$ and let $\mathcal{G}_\omega = \{\mathbf{x}^{\alpha_i} - \mathbf{x}^{\beta_i} : i = 1, \ldots, t\}$ be the reduced Gröbner basis of $I_{\mathcal{A}}$ with respect to ω. Let $\mathcal{K}_\omega$

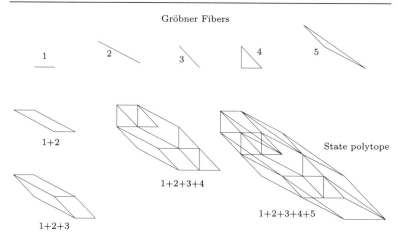

Gröbner Fibers

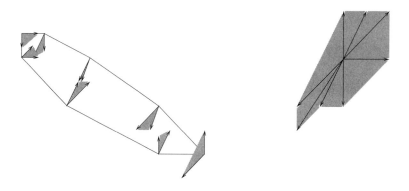

Figure 4. The state polytope of our example projected into $\mathbb{R}^2$.

Figure 5. The state polytope and Gröbner fan.

be the Gröbner cone of $\mathcal{G}_\omega$, $\mathcal{N}(St(I_\mathcal{A}))$ the inner normal fan of $St(I_\mathcal{A})$, and $\mathcal{N}(\text{face}_{-\omega}(St(I_\mathcal{A})))$ the inner normal cone of $St(I_\mathcal{A})$ at the face that minimizes ω. We need to show that $\mathcal{K}_\omega = \mathcal{N}(\text{face}_{-\omega}(St(I_\mathcal{A})))$. The proof is going to need the three facts that we list below without proof. For details, see [**Stu96**].

(1) **Fact (1):** The inner normal cone of a Minkowski sum polytope containing a vector ω' is the intersection of all the inner

normal cones of the summands containing ω'. In particular, the inner normal fan of the Minkowski sum polytope is the *common refinement* of the inner normal fans of the summands.

(2) **Fact (2):** If $x^\alpha - x^\beta$ is an element of the reduced Gröbner basis $\mathcal{G}_\omega$ of I_A, then $\omega \cdot \gamma > \omega \cdot \beta$ for all $\gamma \in \deg^{-1}(A\beta)$. In particular, β is a vertex of $\text{conv}(\deg^{-1}(A\beta))$ and the inner normal cone of $\text{conv}(\deg^{-1}(A\beta))$ containing ω is the inner normal cone at the vertex β.

(3) **Fact (3):** For each degree $\mathbf{b}$ and a reduced Gröbner basis $\mathcal{G}_\omega$, there is a unique element $\mathbf{m_b} \in \deg^{-1}(\mathbf{b})$ such that $x^{\mathbf{m_b}}$ is the normal form with respect to $\mathcal{G}_\omega$ of any monomial $x^{\mathbf{m}}$ such that $\mathbf{m} \in \deg^{-1}(\mathbf{b})$. In particular, if $x^\alpha - x^\beta \in \mathcal{G}_\omega$, then x^β is this unique normal form with respect to $\mathcal{G}_\omega$ of all monomials of the same degree.

We first prove that

$$\mathcal{N}(\text{face}_{-\omega}(St(I_A))) \subseteq \mathcal{K}_\omega.$$

Pick $\omega' \in \mathcal{N}(\text{face}_{-\omega}(St(I_A)))$. Then by Fact (1),

$$\omega' \in \mathcal{N}(\text{face}_{-\omega}(\text{conv}(\deg^{-1}(\mathbf{b}))))$$

for each Gröbner degree $\mathbf{b}$ of I_A. Consider the Gröbner fiber

$$\text{conv}(\deg^{-1}(\mathbf{b} = A\alpha))$$

of $x^\alpha - x^\beta \in \mathcal{G}_\omega$. By Fact (2), $\mathcal{N}(\text{face}_{-\omega}(\text{conv}(\deg^{-1}(\mathbf{b}))))$ is the inner normal cone of $\text{conv}(\deg^{-1}(\mathbf{b}))$ at the vertex β. Since ω' lies in this inner normal cone, we have that $\alpha \cdot \omega' > \beta \cdot \omega'$. Since this is true for all elements of $\mathcal{G}_\omega$, we get that $\omega' \in \mathcal{K}_\omega$.

Now let's prove the opposite containment:

$$\mathcal{K}_\omega \subseteq \mathcal{N}(\text{face}_{-\omega}(St(I_A))).$$

Pick ω' in the interior of $\mathcal{K}_\omega$. Then we know that $\mathcal{G}_\omega = \mathcal{G}_{\omega'}$. By Fact (3), both ω and ω' are optimized at the same lattice point in $\text{conv}(\deg^{-1}(\mathbf{b}))$ for every $\mathbf{b}$ of the form $A\mathbf{v}, \mathbf{v} \in \mathbb{N}^n$. Therefore, ω and ω' lie in the same inner normal cone in the convex hull of all Gröbner fibers. Then Fact (1) implies that both vectors lie in the same inner

normal cone of $St(I_A)$, which proves that $\omega' \in \mathcal{N}(\text{face}_{-\omega}(St(I_A)))$.
$\qquad\square$

Exercise 14.15. Work out the state polytope and Gröbner fan of the configuration consisting of the columns of

$$\begin{pmatrix} 1 & 1 & 1 & 1 & 1 \\ 0 & 1 & 2 & 1 & 0 \\ 0 & 0 & 1 & 2 & 1 \end{pmatrix}.$$

Definition 14.16. A polyhedral fan $\mathcal{F}$ is said to **refine** another fan $\mathcal{F}'$ if $\mathcal{F}$ refines $\mathcal{F}'$ as cone complexes. The fan $\mathcal{F}'$ is a **coarsening** of $\mathcal{F}$. Recall that the common refinement of a collection of polyhedral fans is the new fan obtained as the multi-intersection of all cones involved.

Notice that the Gröbner fan of our running example is a refinement of the secondary fan of the same configuration.

Theorem 14.17. *The Gröbner fan of I_A is a refinement of the secondary fan of A.*

Proof. This is a direct consequence of the fact that the radical of the initial ideal $\text{in}_\omega(I_A)$ is the Stanley-Reisner ideal of the regular triangulation Δ_ω. $\qquad\square$

Exercise 14.18. Verify Theorem 14.17 in Exercise 14.15.

Bibliography

[AL94]   W. W. Adams and P. Loustaunau, *An Introduction to Gröbner Bases*, American Mathematical Society, 1994.

[BFS90]   L. J. Billera, P. Filliman, and B. Sturmfels, *Constructions and complexity of secondary polytopes*, Advances in Mathematics **83** (1990), 155–179.

[BM98]   D. Bayer and I. Morrison, *Gröbner bases and geometric invariant theory I*, Journal of Symbolic Computation **6** (1998), 209–217.

[Buc65]   B. Buchberger, *On finding a vector space basis of the residue class ring modulo a zero dimensional polynomial ideal*, Ph.D. thesis, University of Innsbruck, Austria, 1965.

[CL]   T. Christof and A. Loebel, *PORTA*, available at http://www. iwr.uni-heidelberg.de/groups/comopt/software/PORTA.

[CLO92]   D. Cox, J. Little, and D. O'Shea, *Ideals, Varieties and Algorithms*, Springer-Verlag, New York, 1992.

[CLO98]   ――――, *Using Algebraic Geometry*, Springer-Verlag, New York, 1998, Second edition.

[DF91]   D. Dummit and R. Foote, *Abstract Algebra*, Prentice-Hall, New Jersey, 1991.

[DRS]   J. Deloera, J. Rambau, and F. Santos, *Triangulations of Point Sets*, forthcoming book.

[GKZ94]   I. M. Gel'fand, M. Kapranov, and A. Zelevinsky, *Multidimensional Determinants, Discriminants and Resultants*, Birkhäuser, Boston, 1994.

[GP02]   G-M. Greuel and G. Pfister, *A Singular Introduction to Commutative Algebra*, Springer, 2002.

[Grü03]   B. Grünbaum, *Convex Polytopes*, Springer GTM, New York, 2003, Second Edition.

[HRGZ97]   M. Henk, J. Richter-Gebert, and G.M. Ziegler, *Basic properties of convex polytopes*, pp. 243–270, CRC Press Ser. Discrete Math. Appl., Boca Raton, Florida, 1997.

[Jena]   A. N. Jensen, *CaTS, a software package for computing state polytopes of toric ideals*, available at http://www.soopadoopa. dk/anders/cats/cats.html.

[Jenb]   _____ , *A non-regular Gröbner fan*, math.CO/0501352.

[KR00]   M. Kreuzer and L. Robbiano, *Computational Commutative Algebra I*, Springer, Berlin, 2000.

[Lee91]   C. Lee, *Regular triangulations of convex polytopes*, Applied Geometry and Discrete Mathematics - The Victor Klee Festschrift (P. Gritzmann and B. Sturmfels, eds.), vol. 4, AMS, Dimacs Series, Providence, R.I., 1991, pp. 443–456.

[MR88]   T. Mora and L. Robbiano, *The Gröbner fan of an ideal*, Journal of Symbolic Computation **6** (1988), 183–208.

[Ram]   J. Rambau, *TOPCOM, a package for computing triangulations of point configurations and oriented matroids*, available at http://www.zib.de/rambau/TOPCOM/.

[SST02]   M. Stillman, B. Sturmfels, and R.R. Thomas, *Algorithms for the toric Hilbert scheme*, Computations in Algebraic Geometry with Macaulay 2 (D. Eisenbud, D. Grayson, M. Stillman, and B. Sturmfels, eds.), Algorithms and Computation in Mathematics, Vol 8, Springer-Verlag, New York, 2002, pp. 179–213.

[Stu88]   B. Sturmfels, *Some applications of affine Gale diagrams to polytopes with few vertices*, SIAM J. Discrete Math. **1** (1988), 121–133.

[Stu91]   _____ , *Gröbner bases of toric varieties*, Tôhoku Math. Journal **43** (1991), 249–261.

[Stu96]   B. Sturmfels, *Gröbner Bases and Convex Polytopes*, University Lecture Series, vol. 8, American Mathematical Society, Providence, RI, 1996. MR 97b:13034

[Stu02]   B. Sturmfels, *Solving Systems of Polynomial Equations*, vol. 97, American Mathematical Society, Providence, RI, 2002.

[Zie95]   G.M. Ziegler, *Lectures on Polytopes*, Graduate Texts in Mathematics, vol. 152, Springer-Verlag, New York, 1995.

Index